IDEOLOGY OF RELIGIONS
Scientific Proof Of Existence Of "God":
The Catalog Of Human Population

By Andrey Davydov and Olga Skorbatyuk

HPA Press

ISBN-10: 0988648598
ISBN-13: 978-0-9886485-9-3

Andrey Davydov and Olga Skorbatyuk
Ideology Of Religions. Scientific Proof Of Existence Of "God": The Catalog Of Human Population

Translated by Kate Bazilevsky.

© 2014 Andrey Davydov, Olga Skorbatyuk
Translation © 2014 Kate Bazilevsky
© 2014 HPA Press
All rights reserved.

Table of Contents

Preface. p. v

Part 1. Why People Do Not Believe In "God" p. 1

Part 2. Justifications For New Ideology Of Religions p. 9

Part 3. Basis Of New Ideology Of Religions—A Book That Proves The Existence Of Creator p. 23

Part 4. What Is Human Psyche ("Soul") And What Is Its Structure? p. 31

Part 5. The "Catalog of Human Souls" Exists. Individual Subtype Programs And Manipulation Modes Of *Homo Sapiens* p. 41

Part 6. Dispel Of The Myth About Incomprehensibility Of A Human And His "Soul" p. 57

Part 7. "In The Beginning There Was An Image." What Are Natural Images? p. 65

Part 8. "The Way To Spirit" Is To Study And To Learn p. 73

Part 9. What Are Artificial Images? p. 83

Part 10. "Dead Souls": Human Mutants p. 91

    A Mutant: A Living Human With A Dead "Soul" p. 94

    "Demon's Head" p. 98

    Sleep That Is Death p. 102

    "Affairs Of The Dead" And Affairs Of The Living. Resurrection Of "Soul" Is Possible p. 106

    Sleep Of The "Soul," Not The Mind Produces Monsters. Cadaveric Stains Of Sins p. 112

    To "Be Yourself" Means To Be A Corpse? p. 114

    Not "Human And Nature," But "Nature And Human." An Elephant And A Pug. p. 116

    Genius Perverts p. 117

    "Ye Shall Know Them By Their Fruits" p. 119

Part 11. Barbie Doll And Superman—Sodom And Gomorrah Of Today p. 123

Part 12. "Paradise" Is Here. "God" Is Near. p. 129

Part 13. Responsibility Of Religious Institutions To The Civilization p. 137

Part 14. "Memento Mori": "Soul" Of A Human Is Not Second Hand p. 143

Epilogue p. 157

Connect With Us p. 161

About Us p. 163

PREFACE

We would like readers to know right away our position as authors of this text.

We are not offering any new religion[1]. We are not religious people. We are not people of faith. We do not believe in anything because we prefer not to believe, but to know. We do not insist on anything, or call to any action. We do not express our personal opinion in this book; we presented only objective information as results of our research. And, we, as researchers, have nothing to do with religion or religious institutions. Our field is only research, and tools used in this research do not go beyond a strict scientific framework. We are researchers of ancient books.

The main goal of this book is to acquaint readers with an interesting discovery that we made: one of the oldest books known in this civilization contains, as it turned out, information about nature and structure of the human psyche. As researchers of ancient books, we were able to decrypt some of them. We found about 300 descriptions of human characters in one of these books, and we hypothesized that this book is the catalog of the human population. This hypothesis was confirmed after more than 25 years of testing using strictly scientific methods. And, we would like humanity to know about this discovery, and know how to use the information found in this ancient book to solve practical problems in their daily lives. And, first of all, to provide an answer to the question: "What is the human psyche and what is its structure?" since this question does not have a clear answer in civilization. For those who are interested in knowing scientific definitions of the Catalog of Human Souls (Catalog of human population) and human psyche ("soul"): "The Catalog of human population is a description of type Human by subtype structures. Subtype structure ("psyche", "soul") is a combination of individual archetypes, recorded at the genetic level (principle). Expressions and interaction of subtype structures in manipulation modes and phenological algorithms are described with adjustments for sex, age and cultural differences. Information is recorded on six factors." This definition was developed and introduced by Andrey Davydov—the author of the discovery and decryption of the Catalog of human population ("Catalog of Human Souls").

[1] By the word 'religion' we assume religion as such, i.e. any religious movement and direction that exists in this civilization. By the word 'church' we mean—church as a place of divine services, regardless of beliefs.

In the case of the Ideology of Religions, we tried to tell about this within the framework of a religious worldview; although we could have done it from a position of any other worldview. However, since the Catalog of human population also has another name—"Catalog of Human Souls"—we felt we should appeal to religion because religion is directly related to psyche ("soul"). Solely for this reason, we used the appropriate language—the language of religion. Ideology of Religions that we created is just one of the angles that allows one to look at our research product, and the sphere of activities of religious institutions is just one of the areas where the Catalog of human population can be applied (since the Catalog can be applied in all areas where a 'human' is present). As we see it, even in areas where traditionally a human is not really taken into account; for example, in technological spheres.

To satisfy the curiosity of those interested in whether we have practical experience in creation of ideologies, and on what level—we can provide a couple of examples. One of ideologies titled Ideology of Monarchy was created by Andrey Davydov (the author of the discovery of the Catalog of human population) about 15 years ago, by order of the Head of the Russian Imperial House, Her Imperial Highness Grand Duchess Maria Vladimirovna. Another ideology titled Concepts of Terrorism, created in the beginning of the XXI century, became the foundation of activities of the Anti-Terrorism Center (ATC) of the Russian Federation. It may be of interest that the author of the Concepts of Terrorism is also Andrey Davydov, the author of the discovery of the Catalog of human population. By far not everyone knows this fact. Today, we are also engaged in the development of an ideology, but this field does not apply to religion and religious institutions.

*****

Also we would like to draw readers' attention to the fact that we are not the authors of the "Catalog of Human Souls." We are researchers of ancient books. In this case, we are just translators of the text of one of these books that turned out to be the Catalog of human population. Our work is translation of text from this book within a strictly scientific framework, without introduction of subjectivity, and interpretation, and nothing more than this.

The curiosity of those readers of our books who would also like to know more about us as researchers of the Catalog of human population can be satisfied by a book written by Kate Bazilevsky, which is devoted specifically to this topic. It is called Shan Hai Jing—A Book Covered With Blood. The Story Of Developers Of The Catalog Of Human Population. Despite the fact that this book mostly contains the story about fates of the discoverer of the Catalog of human population Andrey Davydov and his research partner Olga Skorbatyuk, and does not mention those people whose fates were related to research of Shan Hai Jing as the Catalog human population, and who paid for it with their lives, we think that the story of discovery of the Catalog of human population will be of interest not only to those who are already familiar with this source of knowledge, but also to those who have not used it yet and therefore doubt its existence. Since the latter will at least have an opportunity to acquaint themselves with other people's opinion about the "Catalog of Human Souls," opinions of those people who are personally familiar with information from the Catalog and who gladly use it. However, those who doubt the existence of the Catalog of human population, of course, keep the right to consider the following people complete idiots, as they took some horoscope for the Catalog of human population: employees of scientific institutions of Russian Federation such as the Russian Academy of Sciences, Institute of the Far East, for example, well-known scholars of Oriental Studies throughout the world such as Doctor of Historical Sciences, Professor Vladimir V. Malyavin, or Doctor of Philosophy, Professor Anatoly E. Lukyanov, very good specialists in the study of ancient texts, and so on, the highest ranks and analysts of the Russian Federal Security Service, employees of U.S. Department of Homeland Security, who provided us with political asylum, employees of FBI, etc.

However, do note that all individuals and organizations listed above not only got acquainted with a brief introduction to the Catalog, like readers who take the stand "This is a horoscope", "This is another utopia" and "I decided that this cannot be," but also checked this information. Meaning: they tested it in practice. And, only for that reason alone they have concluded that the Catalog of human population is not a utopia or a horoscope. They did not do it on the basis of "it seems to me," as no serious business, no serious researcher, particularly in the scientific environment can afford such, to put it mildly, "liberal approach." Especially since in order to make sure that the Catalog of human population is the catalog of human population there is no need to be a "genius" or have access to the decryption technology. Verification can be done very easily and is available to any person: one can either take descriptions from the Catalog of the human population (free demos are available at http://www.humanpopulationacademy.org/human-programs-demo/) and compare them with real people, their individual qualities, life algorithms, etc., or one can use information about how this or that person can be controlled, apply it in practice, see the reactions and get real results from manipulation. Also, without a doubt, the educational level of a person plays a significant role; and, especially if it is none existent, as it is known, even the most clear and detailed explanation of the issue cannot raise it, as this process usually occurs within higher education institutions.

As it is known, Jesus Christ did not write books. And, from our point of view, he did the right thing because it is a significant loss of personal time. However, Jesus engaged in a different kind of unrewarding activity—preaching. That is: in that what we, who must fully decrypt the ancient treatise Shan Hai Jing, do not have an opportunity to engage in. Unfortunately, we have neither the time nor the possibility to personally educate the masses about the existence of the Catalog human population. The reason is the enormous amount of research that we have yet to do, and, first of all, further decryption of the ancient source that turned out the Catalog of human population.

Simple arithmetic calculations showed that in order to tell each one of more than 7 billion inhabitants of the earth about the "Catalog of Human Souls" for just five minutes—more than 66,000 years are needed. Even with work on weekends, and without breaks for food, sleep and general living. Therefore, we decided that explanation of this subject, at least briefly, but in writing, would "cost" us much cheaper. It is very likely that the amazing person named Lao Tzu, the author of the famous treatise Tao Te Ching (which according to our data is one of the books-commentaries to the Book of Mountains and Seas), reasoned in a similar way when, according to the story, he stopped as he was moving away in an unknown direction, and satisfied the request of the head of the border post Yin Xi (Guan Yin-Tzu).

To those who are interested in our other books—to date we are authors of more than 300 published books. Most of these books (about 290) are descriptions of individual programs and manipulation modes of *Homo sapiens* from the Catalog of human population. Also, among our books are four scientific monographs, which are chapters of a coming textbook on Non-Traditional Psychoanalysis as a scientific direction, based on decryption of Shan Hai Jing. The titles of these monographs are: From Carl Gustav Jung's Archetypes Of The Collective Unconscious To Individual Archetypal Patterns; Archetype Semantics. How It Corresponds To The Concept Of An 'Image': How Archetypal Are Images; Society As A Community Of Manipulators And Their Subjects; Can Archetypal Images Contain Chimeras. In addition, there are four books with complete descriptions of four subtype structures of type *Homo sapiens* in the form on an individual program and three manipulation modes to each program. Despite the fact that these books present such serious (from the point of view of practical application) information, they are written in a simple, popular style. The title of this book series is Manipulative Games For Women: Instruction For Exploitation Of Men. Many of these books can be purchased in electronic format from Smashwords online distributor (see the author's profile—Andrey Davydov (CatalogOfHumanPopulation). Very likely, it is also still possible to find some of the books from the series Manipulative Games For Women: Instruction For Exploitation Of Men (2005 edition) in print via Amazon. A few of our books that were also written in a simple, popular language on issues of gender relations (in other words, relationships between men and women) are available via Smashwords as well: How Men Turn Women Into Nothing; World History Of Turning Women Into Mats; Women's Thirst For Power Over Men—A Way To Utilization; etc. Also, there is a book on the subject of creation of artificial images in regard to parents-children relationships titled Ahnenerbe—Your Killer Is Under Your Skin, as well as articles about the Catalog, such as: Shan Hai Jing: Myths Or Structure Of Psyche; Human Manipulation Modes: Either You Are A Manipulator Or You Are Manipulated. And so on. (See http://www.humanpopulationacademy.org/publications/)

In the text of the Ideology of Religions we tried to explain the key questions concerning the "Catalog of Human Souls." Unfortunately, it is not possible to do it fully in a format of a brief overview. From our standpoint, regardless of how civilization evaluates this information, we had to share it. And, do it in the most popular, accessible form, directly, openly, honestly and without embellishment. We tried to not introduce subjective judgments, personal opinions into this work. We presented the results of our research, presented facts and tried to present justifications and confirmations to these facts.

However, we do not want to be treated like gurus on the basis that we shared this information. We do not claim the position of teachers. From our point of view, this would mean that we already know it all, are perfect, and do not need to study anything anymore. We are more comfortable with a position of students, who still have a lot to learn. We consider the position of students more adequate and reasonable for us.

*****

We are happy to get readers' opinions. However, we are not asking for evaluations of the information that was presented. Anyone's evaluations are not interesting to us. By the word 'opinion', we mean that our opponent has familiarized himself or herself with the offered information in its entirety, analyzed it, drew conclusions, and can support his or her opinion with specific facts. We are not interested in any other types of opinions. Since we, on our part, care to support that what we state, we therefore think we have the right to expect that our opponents will also provide arguments in favor of anything that they state. As far as judgments: since we are not the authors of information offered in this overview, but are only sharing it, we see no point in getting someone else's ratings. We do not think that we or anyone else has the right to and is capable of judging that what has been done by our Creator. According with famous phrase from Tao Te Ching: "Those who speak, do not act"—just talking about the Catalog of human population does not interest us. We are more interested in spending this time on further research of the Catalog and other ancient books, the information that concerns us personally, our lives, our goals, and not on study of opinions of others. We ask to try to not involve us in such processes. We have done more than enough talking about the Catalog for over 20 years: in regular communications, in the scientific community, in media outlets, and so on. By and large, quite a wide range of reactions to this also have been identified long ago and they are predictable. Opinions of people of this civilization are "like copies" because these opinions are not products of intellectual activity of subjects, but are products of their "artificial operating system." They are too boring for us. We have already studied this question in great detail. Therefore, we ask not to be surprised that we prefer to speak with any person about the subject matter specifically, to discuss that specific information from the Catalog, which he has already received and is mastering.

*****

As already mentioned, we are not gurus. And, we do not wish to be. We have a completely different mission, goals and objectives. We think that it is none of our business what readers will do with the received information. We have done our part as researchers: the "Catalog of Human Souls" has been found, decrypted, and returned to civilization. At the cost of our own health, we have made it available to the masses. Religious institutions, and not only they, are notified about the existence of the Catalog of human population. Therefore, the next step is up to civilization. And, it does not matter to us, if it will be religious or any other institutions. It will turn as it will. If religious institutions decide that they do not need the "Catalog of Human Souls," we will respect this choice. The Creator awarded free will to a Human. This free will is not, so to speak, "horizontal" (due to the fact that *Homo sapiens*-bio-robot has a limited selection on how to survive), but "vertical": to seek to "God" as the model, or—to the level of an animal. And, we are obligated to respect any choice. If this civilization decides that it does not need the Catalog of the human population, we are ready to stand aside because we have the proof of existence of the Creator. We do not need to believe that "God" exists—we Know it. And, for us this Knowledge is enough.

Andrey Davydov and Olga Skorbatyuk
San Francisco, California
July 2014

PART 1

WHY PEOPLE DO NOT BELIEVE IN "GOD"

Religion, unlike science, has existed for many centuries. Despite everything, it continues to exist today. From our point of view, this clearly indicates that humanity can exist without science, but not without belief in some "Higher Powers," something supernatural that, unlike modern science, has the ability to solve any problems of humans. One can admit this to himself or not, it does not matter. And, even more so, it does not matter whether this role is filled by biblical "God" (or those gods, which religions offer to consider gods) or not.

However, at some point (it manifested most during the era of the so-called scientific and technological progress) religion in human life greatly receded into the background. Unfortunately, this process continues. It is called the religion crisis. Several theologians and Church hierarchs recognize the weakening influence of religion on society as an existing trend. It was not by accident that Pope Paul VI was forced to admit that "Modern civilization is moving towards an increasingly growing and complete secularization.." Scientific studies also show a progressive withdrawal of believers from church and religion: less than half of respondents answered that they believe in "God," but most of those who admit to being believers actually do not maintain relations with church.

Moreover, according to some of our data, people only declare that they believe in "God." No one really believes in him because human intellect is created in such a way that people are just unable to take something seriously, if there are no reasons and no evidence. To entertain themselves– sure; to play with it as a toy—why not; to dream—no problem, to declare something—sure, but to believe (to really believe)—never. And, unfortunately, this situation relates not only to congregation.

There is another apparent reason to claim that people of this civilization do not believe in "God": their real lives, their actual behavior, their real motivations, goals and objectives. All of the above is either very far from the Creator's "Plan," or even in a completely different range. Then, what faith is there? For example, the belief that fire can burn severely causes people to behave carefully with this natural phenomenon. In regard to our Creator, nothing similar is observed—only declarations. (Below, we present arguments from real lives of people, which do not correspond to the "Will of our Creator", "His Plan" by any parameters.)

If a person really believes, not just declares that he believes, then he acts in line with his beliefs because human beings are simply unable to live and act somehow differently; that is just how they are made. If we draw an analogy with computer technologies, a human is a living machine that works exclusively on programs. Hence, a purely logical conclusion that if people do not live according to that what was prescribed by our Creator, then the words "God's Will" is an empty sound to them, and they do not really believe in "God." If they believed, then the so-called "God's Will" would serve as a single directive, the only, using the computer language, program that they would implement in their daily lives.

From our point of view, departure of humanity from "God" is a very destructive tendency for the human civilization. It is known that prior to today's civilization there were other civilizations that have disappeared at some point; were erased from the face of the earth. From our point of view, their collapse or destruction occurred for only one reason: humanity's loss of Spirituality. And, Spirituality is derived from phenomena such as "Soul" and "God."

As researchers, who consider Spirituality to be the basis of life and Human development, we began to look for deeper causes of the loss of value of religion in the lives of modern people than those that have already been announced by theologians, religious scholars.

And we found the answer.

From our point of view, the main reason is that today religion and church are not able to respond to all of the information requests of modern educated people. And, first of all, to questions like: "Who am I?", "What am I?", "What are the qualities of my personality and what are my natural talents?", "What is the meaning of my existence?", "What is my personal mission?", "How should I live?", "How can I develop?", "What is my personal path to "God"?"

We are talking about not general answers, but answers, which take into account that one person is different from another, answers for each individual.

After all, people expect solutions from any religion, philosophy, ideology, and primarily to issues regarding self-knowledge, self-identity, self-assertion, self-actualization, finding a personal value system, personal philosophy, as well as filling life with meaning, understanding people's motivation, and, of course, recipes on how to live. However, religion and church ceased to be key suppliers of life recipes, key helpers in solving all human problems. While people are used to hoping for at least some support, help from Higher Powers (as more powerful, developed, knowledgeable, able to protect, to help, to teach) that they can get through the "Intermediaries of God."

People tend to lean on religion and philosophy as a source of wisdom known since the ancient times. However, philosophy, from its true purpose (namely a guide with life recipes), turned into a cheap compilation of philosophers' fantasies on the topic of "my love for wisdom." In this sense, religion cannot offer anything either: religion does not have a guide, a manual with personal life recipes for every individual on who should live how and for what.

As for those recipes that religion does offer—as a practical guide, they do not hold water from the point of view of a modern person. Here is why. Human life is multifaceted and full of situational varieties, while the information from sacred books is too general, not specific, and certainly not situational. In addition, recipes offered by religion are for everyone, and do not take into account differences that exist between people.

Here are some simple examples. Religious sources do not have information about what this or that person must eat, if he has a serious disease. And, the same set of nutritional recipes for all people or reliance only on chemical means, as a rule—do not save. One person must eat this, while another person that. Often, this alone can help a person heal. Ancient sources state that food is the best medicine. As for relying on "God's" help, healing by "God": it is practically impossible in the case of a modern educated person, as he or she would rather run to a doctor or a pharmacy. As a result, a disease often remains with a person for life and transitions to chronic. The question is: where to get information about who should eat what and in which cases. However, religious sources do not provide such information.

Also, it is impossible to find information in the sacred books about what professional fields are suitable for every specific person (as it is known, every person has different talents and abilities). A wrong choice usually leads to a number of very negative consequences. Since in nature if you try to make a bird swim like a fish, and a fish fly; or a snake gallop, and a horse crawl—the result will not be positive, neither in regard to their health, nor efficiency and effectiveness of their operations. Practice has shown that it is the same with humans.

There are also no indications of what specific physical activities this or that person must engage in (again, taking into account individual differences); and, not just one activity, but a range. It is known that if there is no physical activity, there is no health; and if there is no health there is really nothing else either. Hence, people need this information.

And, where is information about what wardrobe a particular person should have; what are the styles, materials, and colors, specifically for each person? People are not animals and it is common to wear clothes in human society.

However, for some reason every person feels comfortable and well only in certain styles of clothing, certain materials from which they were made, certain colors, and rejects all others.

There is also no information available on where and how a man and a woman, specific John or Jane, can find not just a sex partner for one night, but a partner with whom he or she can be harmonious, happy their whole lives and raise children together. By the way, in some religions—sex life is considered a sin, a taboo. Which, in turn, not only does not help solve any sexual problems, but also puts a person in a difficult position. From a medical standpoint, lack of sex is either impossible or highly damaging to health. As a result, it is easier for a person to abandon religious requirements and religion as such than to follow such a recipe.

And, where in religious sources is information on how each particular person must develop their intellect? Through which specific actions, study of what specific informational sources? Where, what time of day, with whom, according to what algorithms?

And, where does it say what natural climatic environment is suitable for living and optimal functioning of this or that person? Or, for example, what specific emotional states are normal, usual for a particular person, and in what range and how powerful? And, what are the algorithms of these emotional states? Instead, all that is offered is the same experience for everyone, and emotions that go beyond the "permitted" are advised to be controlled. Although experience shows that very few people can do it because emotions cannot always be controlled by intellect.

And, where in sacred books is information about what specifically a person should do in order to establish contact with another person (personal, friendship, business, professional, sexual)? What specific steps to take, so that another person hears the request or provides the demand guaranteed, listens to the advice, does as required, or what will be better and more appropriate? Or, for example, what to do specifically if someone tries to harm you? To accept? To love your neighbor? Not to kill? "Turn the other cheek?" Alas, in most cases this is either ineffective or completely impossible. After all, at the level of survival instinct—any person wants to survive and protect himself and his loved ones from harm. Plus, situations always require more flexible solutions. And so on. There are many similar examples.

As a result, religious books and religious institutions cannot provide a modern person with individual, thorough, detailed, specific recipes. However, people need to know how to live and how to act. And, they need this information every day in various, ever-changing settings and situations. Yes, of course, religious workers sometimes try to help, but it does not work as it should because their recipe is given without knowledge of what constitutes a person asking for a recipe, what are his qualities, true needs, and abilities. As a result, a recipe that is perfect for one person might not be suitable for and could even be harmful to another.

Even in the case of animals, the situation is much better. A person gets detailed recipes regarding any animal from reference books, encyclopedias, textbooks: how to recognize them, how to distinguish from each other, how to interact with them, how to raise them, what to feed them, how to treat, tame, and hunt them. This occurs only because there are catalogs of animals, and in those catalogs are clear descriptions of particular types. In the human community such descriptions are not presented. Then, what kind of basis for giving advices, recipes from the category of "what to do?" and "how to live?" can even be discussed? On what basis then can one person advise, recommend something to another? On the basis of his or her own ideas? If a person would try to act in such a way with any animal, then it is unlikely that something positive would happen, as either he or the animal would suffer.

Therefore, from our point of view, it is absurd to consider recipes from an adviser, who does not have information about what specific qualities on all factors (intellect, physiology, nutrition, emotions, sexuality, environment) the requestor has. Since, in this case, there is no basis for the issuance of recipes. To continue the analogy, if a person did not have information on how to tell a cow from a crocodile, then having fed silage to a crocodile and some meat to a cow—he would get only two corpses.

Unfortunately, a single recipe for all people does not exist and by definition it cannot exist. This is well illustrated in practice. However, religious institutions do not have the other kind of recipes—the individual ones.

However, criticism of religion is not our goal. We think that science must help religion gain its original status in people's lives: the status of the main source of recipes of life and information about a Human. From our point of view, this is exactly how church can regain its status of the primary social institution. However, more importantly, through religion and church people will be given back core values: "Soul," Spirituality, "God."

Regardless of race, nationality, or religion, every person seeks spirituality, and that is a natural desire. However, it so happened that in our time there is no religion that can give its adepts the recipes mentioned above, including recipes on how they can become individuals equal to the Creator (no matter if it is "God," Yahweh, Mohammed, Buddha, etc.). At the same time, for example, the Bible states: "And God said, "Let us make humankind in our image, according to our likeness" (Gen. 1:26). Then, where are people like him, for example, like Christ? Christ did not claim to be "God"; on the contrary, he said that he was not "God," but a human being, a "Son of God." Then, to be human and a son or a daughter of "God" means to have qualities and capabilities that Jesus demonstrated? After all, synonyms of the word 'likeness' are 'similarity', 'analog', 'equality'. If it is not so, then which of the sources we must not believe: the Bible or the thesaurus?

We are not offering discussions on this topic. For, as practice shows, all intellectual discourses lead only to even more confusion and speculation, while the problem does not get solved. We have attempted to work specifically on the solution to the problem and turned to ancient books. Since, as it is known, there are books much older than the Bible, Torah, Quran, etc. Some of these books date tens of millions of years back. Studying one of these books (according to some researchers, this manuscript dates 21st century BC), we found answers to all of our questions because this book turned out to be nothing other than the "Catalog of Human Souls." It is a catalog because this book contains descriptions of about three hundred models of psyche of *Homo sapiens*. Moreover, it holds detailed instructions on how this or that person can realize the main "Divine Plan": make himself or herself "the image and likeness of God." In this book—about three hundred ways to achieve this state for humans were discovered.

It turned out that "God"-Creator has taken care of the "Crown of His Creation" and left humanity a detailed instruction in the form of about three hundred descriptions on who is who, who has which innate qualities of psyche and physiology, character traits, life algorithms, functions, talents, etc. And, also left to the children recipes of life on all six factors: intellectual, physical, nutritional, emotional, sexual and environmental. That is, in this book, "God" lovingly left information on what each specific person should do: how to live, how to behave in certain situations, what to do, who to be, and, notably, how a human being can become like the "Father." Paradoxical as it sounds, it is not difficult to accomplish, there is no need to, for example, sit facing a wall for 10 years, to torture or strain yourself, or to limit yourself in something. All that is necessary is to study. However, first, it is necessary to have recipes, step-by-step instructions on how to do it.

You will be able to get acquainted with information about the "Catalog of Human Souls" in more detail below. From our perspective, what is most important is that this book is a direct proof of existence of "God"; the segment which religious institutions and their congregations have always needed, but lacked.

PART 2

JUSTIFICATIONS FOR NEW IDEOLOGY OF RELIGIONS

Unfortunately, it should be noted that even today, in the enlightened XXI century, "God" remains an unknown value to humans. Although "God" is the fundamental value in this world, as he is the Creator of all: of us and everything else that we see around us.

Also, it is regrettable that a human being, who according to religious canons is "the Crown of God's Creation," is still far from known. The world around us has been studied much better than the being called 'human'. Proof of that is the fact that this civilization has created encyclopedias and directories of anything: plants, animals, minerals, vehicles and any other technologies, clothes, shoes, art and so on; however, there is no catalog of human population. This refers to a catalog as a source of objective information, created by analogy with, for example, an encyclopedia of animals, where it was possible to easily and quickly obtain information about each person, to know who he is, how he lives, what his aspirations are, what is the motivational basis of his life, and how to interact with him. Since such a catalog does not exist, knowledge about the main "God's Creation" is incomplete because a complete knowledge of a subject suggests the possibility of a detailed description of properties of this subject.

Of course, human physiology has been studied and, in a sense, it is known (although not fully, according to confessions by medicine and other sciences about the human body). As for psyche, or, to define more precisely—the human "soul," science has "put its hands down" on this issue a long time ago, and accepted failure. Psychologists and other experts in the field of psyche felt helpless, and therefore were compelled to admit the "soul" to be in fact non-existent. They began to consider the nervous system, brain activity, and intellect as psyche; although the "soul" and psyche are, in fact, one and the same. And, intellect is not psyche, but only a part of psyche. And, not the main part, as it is commonly assumed.

Then, what kind of a detailed description of *Homo s.* can there be, if a significant component ("Soul") is unknown? In the eternal debate of psychologists: what is primary, "soul" or body, "soul loses" only because psychologists do not know anything about it. Despite the fact that "soul" (psyche) is the focus of this "science."

Shortfalls of medicine, psychology, anthropology, and other sciences that study human beings affected people's attitude towards religion. The idea that the "Soul" does not exist, and therefore, religion is not necessary has been ingrained in people directly or indirectly. Religion was transformed into anachronism, or a declared topic of interest of under-educated, poorly informed people.

Another problem is that religion cannot provide definite, complete, and, most importantly, well-reasoned answers to questions about those phenomena that are directly related to the field of religion. Even the basic religious concepts, "Soul" and "God" (Creator), are not specific, are too vague. While people who live in the XXI century are very different from people who lived during past eras. And, the vast majority of modern educated people are not ready to believe in that what does not have specifics, evidence and wide practical use.

People are willing to believe only that what one can see with his/her own eyes, test, and, of course, use in everyday life. That is normal, natural. However, the current state of affairs does not in any way contribute to appearance and strengthening of sincere "Faith in God" in the hearts of people. People cannot believe in that what has no definition and no description. They are only able to declare that they believe.

To this day, religions try to communicate understanding and recognition of the Creator to people; understanding of the role and purpose of His Creation, the importance and unique mission of every living creature, and, first of all, a Human. However, how can this be done in the absence of basic information inside the religious confessions themselves? A common definition of "God" does not exist. (As well as a clear, specific, common definition of 'religion'.) As for the definition of who is a Human—according to our data, incorrect information is presented in this regard.

Certainly, religious, philosophical, mystical, scientific arenas try to offer answers to questions "Who is "God"?" and "Who is a Human?" However, the process is very unproductive because in their views they contradict both each other and logic. Such ambiguity, discrepancy, contradiction, confusion causes rejection in any person who is capable of observing, thinking, and analyzing. Any phenomena can have only one meaning, one value, one core, and therefore, can have only a single definition.

Also, from our point of view, a modern human cannot accept the idea of the existence of "God" due to lack of evidence that "God" exists. This is quite natural for a human as a potentially sensible being. A Human was created to be sensible. Therefore, expectations that people, especially modern, will believe in that what has no evidence are naïve, to say the least. In order to perceive the idea that "God" exists and that "God" created a human, one must have proof of the "Divine Act" of Creation.

However, serious, undeniable evidence for this never existed in this civilization. As a result, scientists create multiple theories, one more absurd than the next, about how this planet was formed, how the Solar System was created, and how a human being fits in all of this space. Again, there are many theories, but there is only one truth. And, the truth is that "God" (Creator) created all this. It is also true that without serious arguments in favor of this, people are not ready to believe, to accept this idea. And, faith has never replaced Knowledge.

Specifically due to lack of knowledge and understanding of such phenomena as "God" and a Human, as well as real evidence, many spiritual movements, including religious invite only to believe. This only leads to the fact that they are continuing to lose their positions in modern social life. And, if this trend continues, the existence of religion as such, as well as religious institutions, and especially their impact, will end sooner or later.

We suppose that it is necessary to do everything to ensure that this never happens. Therefore, we propose a new ideology of religions and its practical implementation. We propose to finally determine the answer to who is, at least, a Human. And, to move the theory of existence of "God," the human "Soul" from the status of a theory, which can only be believed in, into the status of a theory that any interested person can test. Of course, also providing proof that "God" really exists.

In our opinion, in order for people to be able to live a spiritual life, the Church must become the main place where they can get information about themselves, other people, answers to their questions, recipes. In other words, where they can obtain information about the "Soul," as knowledge about the "Soul" contains information about everything listed above.

Then, Church will have a primary role in people's lives. Note that church did not play a primary role in the lives of people in previous centuries. Yes, its significance in people's lives used to be much higher. However, in addition to secular social institutions that have always shared "power over the minds" with the church, another significant competitor was magic. Healers, sorcerers, shamans had a huge recipe base, along with great power and influence on the human society. As it is documented in numerous sources, they helped people solve their problems in reality; for example, health issues.

A new ideology of religions, created on the basis of evidence for "God's" existence produced strictly by scientific methods, can bring people back to "God."

*****

A human needs "God." As already mentioned, from our point of view, the source of all human misfortunes, sufferings, and problems is the lack of true

Spiritual Life. And, its absence is a consequence of absence of specific, authentic information about what "Soul" is. The presence of contradictory information in this regard leads to a conclusion by human intellect that "Soul" is a phantom. And, if "Soul" does not exist, then what is Spirituality? After all, the concept "Spiritual" is a derivative of the concept of "Soul."

The crisis in this direction worsened so much that it reached a point when even priests began to allow simply formal engagement with Spirituality. From our point of view, formal attitude towards religion takes place in a case when a person who fulfills all religious rituals, does not consider "God" as an example, as a model for his or her personal life, for his or her own behavior, for mimicking. But then, in this case, where is the fulfillment of the "Divine Will": "And God said, "Let us make humanity in our image, according to our likeness?"

People perceive "God" in a variety of ways: as a benefactor, who must continuously give something to people; as a punitive authority; as 24-hour surveillance; etc.; but in any case, not as a creature to take an example from, to imitate, to follow. People love to speculate about merging with "God," but in real life and in their consciousness they always divide "God" and themselves. And, consider "God" as "Santa Claus," in whom they want to believe, although it is a fairy tale. And, most importantly, supposedly "God" is irrelevant in their real lives.

The result of this position is presented and well illustrated by people's daily lives. Even those who call themselves believers lie, cheat, act meanly, are lazy, do not want to work and learn, fight, transgress the moral laws, harm themselves, people around them, as well as nature. Some even show overt violence—they destroy, rob and kill. Would this happen if people perceived "God" as the only example, a point of reference in their lives? Or, would they at least try to be like him?

*****

While engaged in analysis of the situation in the sphere of the spiritual life of this civilization, we noticed a number of problems, which, from our point of view, can be addressed by introducing a new ideology of religions.

The main problem has already been mentioned above: none of the existing religious and philosophical-religious trends and directions provide an opportunity to get answers to questions like "Who am I?", "What am I?", "What was I born for?" to any person. They cannot provide strictly individual, personalized recipes to every human. As already mentioned, at the present time the only help that church offers is mostly emotional support to sufferers. However, an opportunity to be comforted in sorrow, sadness, and anxiety is certainly good to have, but this does not provide information or recipes. Compassion does not replace information.

Emotions do not replace purely intellectual needs of a human. Therefore, many people simply do not need such help from the church.

People need something else. While in response religion and philosophy offer them to search on their own for their selves, their individual "Paths to God," to Spirituality. To search anywhere: in this or that doctrine, in "God" or "Gods," within their own selves. And, this does not give the desired results. If this technique worked, people would not have any problems with self-knowledge or self-determination, as well as all other aspects of their lives. All people would be healthy, live long, without creating problems for themselves and others, and there would be peace, prosperity, harmony and happiness. However, this is not seen in real life. On the contrary, all of humanity's problems are multiplying and multiplying, like mushrooms after rain.

From our point of view, another problem, based on which a crisis of religion occurred, is the difficulty of use of existing religious recipes in people's daily lives.

For many people this is the basis for rejection of religion as a provider of recipes, which in practice are either insolvent in terms of achievement of desired results, or are just impossible to follow. (An example of this was given above in regard to sex.) Such recipes can hardly be highly appreciated. This is another source of human unwillingness to participate in church life. For the Supreme Being, the Creator, by definition, cannot create non-working schemes, tools, recipes, as well as impossible objectives and requirements. That is contrary to reason and logic, and it is difficult to suspect that our Creator is irrational and illogical. The conclusion made is simple: there is no "God."

Puzzling proposals, existing religious recipes make modern people turn away from the church. And, it is not surprising. Not knowing specifically what to do, a person constantly makes mistakes, which he is forced to pay for, and sometimes they are very expensive. Often, the result is annoyance with religion and church, and sometimes even outright blasphemy, an attempt to blame the Creator for one's troubles and misfortunes, as well as other deliriousness. A human does not want to suffer; he wants to live a good, happy and prosperous life, no matter how much you tell him that "God suffered and told us to." Not finding the recipes on how to achieve this at the church, a human tries to get this information from secular institutions.

*****

Human consciousness and his interests are anthropocentric. For as long as people exist, they primarily want to know everything about themselves, their purpose and problems. According to our research, the desire to know oneself is based purely on pragmatism. It turns out the reason is in that "God" created a human in such a way that he is not able to function effectively, behave correctly without clear knowledge and understanding of what he needs to do, without an instruction, without a clear program of actions.

"God" made representatives of, for example, the animal kingdom much simpler: they are born and exist on the level of living machines with a clear program of their lives inside them; it is possible to say that they have an innate knowledge of what they are and a complete collection of recipes on how to survive in this or that situation. Therefore, animals do not need to think or discover themselves, to study or to improve. However, humans "God" created differently. A human is given an opportunity to think, to learn, to study and to change, to evolve, which is impossible for any animal. (In any case, to evolve not from an ape to a human, as, for example, Charles Darwin mistakenly believed, but to evolve towards his main guide—"God" the Creator.)

Hence, the constant human desire to acquire knowledge, and mainly about his own self. Apparently, the desire for self-knowledge is inbuilt in a human being by the Creator since he did not give people intuitive knowledge of their natural program, as it was done with animals. Therefore, he stimulated their desire for knowledge as a process, as a function of a human being that distinguishes him from an animal. Since it is the natural program that allows both animals and people to survive and to thrive.

Representatives of other biological types in nature clearly demonstrate that the program that is given by "God" to every living creature is the source of his success and survival. If the subtype program of any animal gets wiped out, then it will not know how to behave, how to survive, and, as a result, perish very quickly. It turned out that this principle works in relation to human beings. If a human does not know himself, who he is, or in other words, what is the program that "God" implanted in him, then that person is lost, makes mistakes and sooner or later dies as a result of these errors.

Today, religious institutions, church unfortunately do not provide people knowledge of their natural program. It is no wonder that in search of instructions on how to act people turn to secular specialists, and have more hope for science, technology and secular tools. Although it is quite obvious that non-religious, secular toolkit in many cases does not work. In many cases, modern science is also of little assistance. Especially since at its current stage science is increasingly aimed at solving technological problems, which, unfortunately, are often solved at the expense of human problem solving. Whereas a human is "God's" creation. A human was created through Nature, which was created by "God" the Creator, so it is not a surprise that most human problems cannot be solved by technological progress.

Moreover, technological progress by large suppresses the process of unfolding of human potential. It is possible to unfold this potential, those talents and abilities that "God" awarded to each person, only through knowledge of his natural program, i.e. how "God" created and wants to see him. Untalented, incapable people do not exist in nature—there is only lack of information on which talents a person has and how to develop them. As we know, Jesus Christ did not need any tools of modern communication, modern transportation, medical equipment, computers—nothing of that what is offered by what became a technogenic civilization. But, does a Human really need this? Or, after all, would it be more useful to get instructions on how to become a creature that can easily do without all this technical variety?

*****

We agree with the opinion of the Church that all people must finally become "Disciples of Christ," and not formally, or in their fantasies, but in reality. For "God" made humans "in His image and likeness," and Christ, "His Son," convincingly demonstrated by his own example what sons and daughters of "God" (people) should be like. However, unfortunately, Christ did not leave people the methodology on how he reached supernatural (divine) abilities, which he possessed. We mean that the "Son of God" did not leave textbooks, instructions detailing the steps on how to become like him; that what is left are only stories about him and his teachings, passed on orally. This is very unfortunate because for this reason second Christ did not appear in the human society.

The Church seeks to communicate to people the idea that their purpose in life, their "Path" must be Evolution. And, that this evolution must take place due to a one's own efforts. And, this is right. For evolution, as a qualitative change in personality, its level, its capabilities, is the essence and the result of a human's Spiritual Path. However, in practice, religious institutions do not have specific, detailed, personal methods for every person who wants to evolve.

Evolution as a mechanism is innate in each person; this is another purely human characteristic, absent in other biological types. And, people would be happy to do it, but they are offered ways and recipes of self-improvement that are common for all. Although it is obvious that everyone is different and for evolution to take place, individual personal recipes are required. Otherwise, as almost 2,000 years of history have shown, there was no one among people equal to Christ, the "Son of God," by qualities and abilities, and there will be none. And, this situation contradicts the will of "God," who said that He creates a human in his own image and likeness, and thus made it clear that he expects human beings to match Him. Actually, we think that the achievement of Likeness—that is "Unification with God", "Dissolution in

God," and not fantasies based on feelings that people are comfortable with presenting as a process of "Unification with God," while believing themselves to be already equal to "God."

*****

We suppose the church must have a method of providing spiritual growth in its hands. And, what is no less important, is that this method must be based on a source of knowledge that is not a source of subjective views.

From our perspective, possession of such a method by the Church has a great practical importance since only through Spiritual Growth people can achieve health, prosperity, harmony with the world and with other people, and develop their full potential. "God" created a human differently from all other living creatures, to which a primitive life (food, reproduction, etc.) is enough to thrive and be healthy. In the case of *Homo sapiens*, it is not sufficient to execute according to only these conditions, as "God" expects more of humans. As already mentioned, our research results can help the Church in this matter.

*****

As for solution of another objective of the Church, namely, to make people love each other, live in peace—currently this task is not feasible. It is unfeasible only because people do not have any idea about each other's real needs. In view of this, even when a person sincerely wants to help someone, the maximum that he can offer to another person is the same thing that he would like to get (to have) himself.

However, at best, such method leads to dissatisfaction on the receiver's part, and at its worst causes the receiver real harm instead of good. By analogy, it is the same as if you take a fish out of water and thrown it the air where the birds fly, and submerge a bird in the ocean's depth where the fish live. If you were to use this method on practice in nature—both animals would die. As for the giving side, a similar experience, often even out of sincere caring, leads to a deep disappointment in people, and a lack of desire to help anyone else. Also, such experience often leads to a negative attitude towards other people in general, and, as a consequence, to aloneness. Although in fact, the problem has a very simple solution through knowledge that every person is different, that people have different characters, life algorithms, basic needs, objectives, goals and dreams. But most importantly, through specific knowledge of who the other person is. Then, it would be easy to offer him only that what he really needs, that what is desirable and useful to him.

However, at present day, people are simply unable to love each other, and this is happening only due to the fact that they absolutely do not understand the other person. And, they do not understand because they have nowhere to get information about him or her. It is impossible to fully know another person intuitively; intuitively it is only possible to trace a tiny fraction of his or her true needs, but even that is not accurate. And, an inaccuracy is already a mistake. For if a person is not familiar with a dog-care guide, but only heard somewhere that dogs love bones, then he will feed the dog chicken bones, from which the animal might die. Therefore, one person trying to judge what the other person is like and what he needs based only on his subjective perceptions is a 99.9% guaranteed error. If people need information from encyclopedias on how to handle animals or plants, an encyclopedia about humans is needed as well. Or, is a human being more primitive than, for example, a dog or a cactus?

*****

Unsolved problems deprive people of the two most important sources of normal, prosperous, happy life, which are: their "Soul" and their "Longing for God." Religious institutions of any direction are very concerned with this situation and they are trying to help people. However, to date, it is extremely difficult to call this process successful. Intoxicated by achievements of secular civilization, humanity is moving farther and farther away from "God," religion, and church. Not finding recipes and answers to their questions within their religion, people often look to other religions and religious cults, sects. Or even become disappointed in religion and believe only in science. Or do not believe in anything except the "Golden Calf."

We think that this is not right. "God" created a human, and he or she should be with "God" all of his life. He or she must follow the way that "God" intended for him or her. And, follow it clearly, without going off the path, and without interruption for any secular temptations. Then, a Human will be saved. We think that salvation of a Human begins with salvation of his "Soul." "Soul" is the foundation of a Human. Spiritual life is the foundation of all human life. However, in order to revive the spiritual life of modern people, a new ideology of religions is needed. Otherwise, religion will not be able to fit in with the modern conditions. And, one day it will simply be removed from life of human civilization.

*****

First of all, from our point of view, church should stop being only a place for worship and a place for conversations with a priest. As practice has already shown, an attempt to overcome the crisis by modernizing the church, does not give the results that were expected. For this reason, it probably makes sense to stop "flirting with people" and trying to involve them in religious life by any means (it has already come a point that services almost turned into discos).

It is not a surprise that in connection with such politics, many parishioners see the Church as a place of entertainment, another show. Parishioners come to church to listen to the preacher, to hear music and singing, to smell the olibanum, to admire images in the temple. In other words, church is perceived as a lecture, a trip to a museum or a concert. Or as another networking event because Church is increasingly becoming mainly a place of leisure, a place to meet people: "I will go to church, talk to people." But where is the most important thing—communication with "God?"

Seeking a way out of the crisis, church is often ready to transform itself into anything. From our point of view, such an organization of church life leads to the profanation of Spirituality. It is no wonder that instead of going to church, the vast majority of people prefer secular lectures, a real museum or an art gallery, a musical show or a concert, or a crowded social networking event. People realize that there is no spirituality in either place, but social events are much more diverse, colorful, interesting, more attractive. And, as far as going to a priest, people prefer to visit a psychoanalyst, a psychotherapist or simply meet with a friend(s) with whom they can talk, get sympathy, support, and advice. We think that this is not right. Lack of true Spiritual Life is extremely detrimental for a human; that is, for an individual and for humanity as a whole. A human is not an animal. A human differs from animals by having a more complex psychological structure, in which Spiritual Life must be present. Spirituality must be the basis of human life.

*****

We could continue to further list problems that are currently present in the arena of religious life. However, it seems to us, that it would not make sense. Religious institutions are well aware of them without us. We would like to contribute to the solution of these problems. And, we are capable of doing it based on research and discoveries that we have made and continue to make.

We are ready to help the Church bring people back to "God" as a beacon in their lives; to religion and to Church as the main helper to believers because we have found real proof of existence of "God," and proof of existence of human "Soul." Due to this discovery people can strengthen and regain their True Faith; no one would even consider doubting that "God" exists, that "God" created human beings and that they should follow Him, the "Divine Way." And, in turn, the church can help people follow this "Way."

Based on our research, and as a result of our research activities, which meet all existing standards of scientific research, we are ready to present people with the answer to the question: "What is the "Soul" of a human, and what is its structure?" Thus, confusing, unclear, and vague definitions of "Soul" and "God" of the past will finally have specific meanings, and through that— become valuable in lives of every human being. For any modern educated person will be able to not just believe, but to know for sure that human "Soul" does exist, and that there is "God." From our point of view, acquisition of such confidence is the best, the firmest foundation for Church (as a guide on the Path of Spiritual Life and Self-Development) to gain primary importance in every person's life.

In this case, there will no longer be a need to exhort and persuade people to believe in "God," to follow his commandments, and to visit the "Temple of God." Entertainment, shows also will no longer be required. Church and churchmen will be really needed by people, as the only helpers and advisers in their lives. The church will no longer have a competitor in the form of science, as currently secular science and secular culture are not able to solve most of the problems that humans have.

PART 3

BASIS OF NEW IDEOLOGY OF RELIGIONS—A BOOK THAT PROVES THE EXISTENCE OF CREATOR

It seems to us, we managed to find the solution to the problem of how to bring people back to "God," to make "God" the only beacon in every person's life. We have found a book that (not only from our point of view, but also based on many objective indicators) is the proof of existence of the Creator, "God," the Great Creator Architect of the Universe, in other words—the Supreme Force.

A Russian researcher, Andrey Davydov has been studying ancient manuscripts since 1975, and has managed to decrypt one very ancient source. Andrey Davydov is a sinologist and research supervisor of a Special Scientific Info-Analytical Laboratory. Back in 1975 he suggested that one of the earliest treatises is nothing other than a description of structures of the human psyche. This marked the beginning of his study of this book as the Catalog of human population. Although at that time it was only a hypothesis.

For nearly 20 years, Andrey Davydov gathered keys to the mysterious code of this manuscript. By the mid-90s of the XX century, scattered elements of the puzzle came together, and research related to the selection of correct keys was over. At that point, the process of decoding of the content of this treatise has begun: translation of descriptions obtained therefrom to human psychophysiology and practical experiments.

As a result of decryption of this ancient manuscript, Andrey Davydov uncovered belongingness of any person to a particular subtype within the type *Homo sapiens* through certain laws of mechanisms and work of psychophysiology. It turned out that each representative of the 'human' biological type has certain stable qualities, which distinguish him, as a representative of one subtype structure, from representatives of other subtype structures.

Extensive knowledge in a variety of scientific disciplines helped Andrey Davydov decrypt the ancient manuscript: sinology, biology, soil science, geography and physical geography, mineralogy, geomorphology, ichthyology, ornithology, ethology, and other natural sciences; as well as his knowledge of cultural studies and mythology of different countries.

As a result of decryption of the ancient book, Andrey Davydov got descriptions of existing qualities of existing real people, and proved that the Catalog of human population exists.

Descriptions, which he presented have been compared to qualities of real, specific people within strict scientific experiments. According to estimates of professional psychologists, descriptions submitted by Mr. Davydov correspond by almost 100% to qualities of subjects, which were easily observable or uncovered using other scientific methods. Psychologists at the MSU (Moscow State University, Moscow), as well as psychologists at the ministries of power structures of the Russian Federation, were forced to admit that the methodology of uncovering of subconscious structure of psyche developed by Andrey Davydov surpasses methods and technologies, which they use to determine personality traits of a person.

Also, experts admitted that prior to experiments with the use of information from the Catalog of human population, not a single precedent existed when a psychologist would provide such an in-depth, detailed description of people whom he did not know, never met, did not communicate with, etc. Thus, Andrey Davydov's research product was verified according to criteria accepted by science. With application of a standard test in the form of direct experimental verifications, Andrey Davydov's hypothesis that the treatise that he researched for over 25 years, is the "Catalog of Human Souls," turned from a hypothesis into a theory that has proof.

Between 1997-2002 Andrey Davydov introduced the scientific world to the discovery of the Catalog of human population. For example, in 1997, there was the First Russian Philosophical Congress "Human Being-Philosophy-Humanism" (Volume VII, The philosophy of human problems, edited by corresponding member RAO L. A. Verbickaya, and by associate professor B. G. Sokolova Saint-Petersburg, 1997). He also spoke at events like round table discussions, scientific conferences (including international) devoted to anthropology, philosophy and so on. For example, in 2002 he presented one of his papers about the discovery of the Catalog of human population in Moscow at the International Conference of Prospects of Preservation and Development of Uniform Planetary Civilization: Culture, Ecology, Cosmos. He also published articles about the Catalog in Russian magazines. For example, for some time his works were published in one of the journals published by Bauer in Moscow. Between 2005-2006, presentations took place on the Russian television, and included a series of talk shows featuring Olga Skorbatyuk (a professional psychologist and Andrey Davydov's research partner), in which she shared this scientific discovery with wide audiences. During the talk shows, Olga Skorbatyuk told about qualities of people born on certain dates, and since the talk shows were live, the audience called into the studio with questions and opinions; many confirmed with surprise that they themselves or someone they known really possess personal qualities, which were being presented during the shows. (A recording of this series is available for review.) Also, in 2005, a series of books was printed in Russian (authors—A. Davydov, O. Skorbatyuk). They contain psychological descriptions of specific people, as a demonstration of possibilities of the technology created by Andrey Davydov—the technology of decryption of an ancient manuscript that turned out to be the Catalog of human population. (Several printed copies of these books are available for review.)

*****

After getting acquainted with the above-mentioned ancient manuscript, in 1975 of the XX century, Andrey Davydov suggested that it is not just some sort of mysterious and incomprehensible description of flora, fauna, chimeras, spirits and deities of the mythical world, but is an encrypted description of people's psyche, their subconscious sphere, or, in other words, human "Soul." About 300 such descriptions were found in this ancient book. Moreover, it was discovered that this book contains descriptions of individual "Paths" of a human to "God," ways to realize the "Divine Plan" in respect a human—for a Human to become "in the image and likeness" of Him.

After many years of research of this manuscript, Andrey Davydov suggested that this book is the proof of "God's" existence because it contains detailed descriptions of people as they really are, without social masks and roles, under which their true personalities, goals, plans, and preferences are usually hidden; that is, as humans were created by the Creator.

Andrey Davydov hypothesized that, being a peculiar "blueprint of humanity," this book cannot be an artifact. This hypothesis has been made on the basis that if the human species is divided into subtypes, and existence of which in this civilization is not known, then the author of this book could be only the Creator.

Also, in the opinion of Mr. Davydov, only the Creator who created a Human could have left such a detailed description of varieties of human "models." These descriptions are similar to instructions to modern technical tools, to the level of detail with which these tools are described: what materials and parts it is made of, how it functions, what are its maintenance and exploitation conditions, what are its fault rectification options, and so on. In civilization, no other source created by people has been found with such descriptions. Humanity cataloged almost all objects visible and even ones not visible without a microscope or a telescope, but could not describe themselves in a similar way, and were unable to catalog themselves. As many thousands of years of practice have demonstrated, no researcher was able to create such a product. If this was not so, humanity would already have the Catalog of human population, created by the people themselves. However, it is missing from civilization.

Another reason for Mr. Davydov's hypothesis that the manuscript that he discovered is a book left by "God" to humanity, was the absence of the author of this manuscript, as well as at least an estimated date of this source, despite the fact that this manuscript has been studied for more than one millennium. For example, mentions of this book are found in writings of Confucius, the ancient Chinese thinker and philosopher, who lived around 551 BC.

*****

The ancient book that turned out to be the "Catalog of Human Souls" is not hidden in some archives or repositories. It is available to the general public. This manuscript (although encrypted) is freely available in bookstores of many countries and in different languages. It is also available in libraries, as well as for online reading. This manuscript is very well known in the circle of specialists, especially those who study ancient books. However, it is famous for incomprehensibility of its content. For this reason, for many thousands of years in a row, neither readers, nor researchers could understand what this book is about. Due to lack of an answer to this question, this manuscript has been classified as an ancient literary monument.

At first glance, this book does not describe anything special: there are mountains, soil types, a variety of animals, grasses, flowers, birds, insects, rocks, streams; very similar to an ordinary landscape. However, on the other hand, on this landscape outlandish (and sometimes strange) creatures run, fly, rush about, and make some actions: unthinkable chimera with multiple heads, arms, legs, tails, and wings, as well as mythical heroes, spirits. Throughout time, in all the countries in the world many researchers of this book were very confused by such mysterious content. All as one were puzzled: what is it? Data on historical geography? A mythological directory? Or is it something else?

Traditionally, this manuscript is considered a work that combines knowledge from different time periods on descriptive geography, landscape studies, zoogeography, botanical geography, mineralogy, ethnogeography, cosmology, medicine, as well as myths, legends and beliefs. There also exists an opinion that this book is an encrypted map of the sky and a calendar system. Also, there is an opinion that it is an outline of the system of the Universe. There are many hypotheses in this regard. Confusion of the researchers is easily explained since the world described in this book is so strange, outlandish that it seems as if the ancient purposely created this work for fun, to ridicule the now living: go ahead, puzzle over this!

Incidentally, this manuscript was not included in the category of Confucian classic books, and very likely for the same reason—due to lack of understanding of its content. On the one hand, the ban that Confucius placed on the study of this book delayed humanity's knowledge of "God's Plan" for itself for many centuries. On the other hand, this ban saved the "Catalog of Human Souls" from any changes and subjective speculations. As a result, the ancient book, which turned out to be the Catalog of human population was preserved in its original form.

Researchers did not come to a consensus not only in regard to content, but also in regard to authorship of this book. Some claim that it is an anonymous literary monument, while information from other sources suggests that it was created by a semi-mythical emperor, who lived in the XXIII century BC; but in fact, the author is still unknown.

According to a legend, an assistant of a certain mythical hero has engraved this treatise on nine sacred vessels. This legend reads that this hero coped with an unseen flooding that struck the earth (perhaps it was the Deluge, but this is only a speculation) and has arranged it. After he has arranged the earth, he acquired knowledge of its rivers, mountains, spirits, as well as animals and plants.

Then, he ordered his assistant to describe all that was seen in detail. These records, along with images of spirits, amazing outlandish animals, birds, and plants were engraved on tripod vessels. Later on, as the story goes, these sacred vessels were lost for some reason. However, prior to their strange disappearance, the text of the "Catalog of Human Souls," together with drawings has been copied.

It is also interesting to note the fact that, despite the enormous amount of research done on this book, even an approximate date of this enigmatic work still does not exist. According to suggestions of some researchers, the date of this treatise is approximately XXI century BC.

*****

It is also interesting that in the mythologies of all cultures of the world (those that continue to exist, as well as those that do not) are facts that indicate that the "Catalog of Human Souls" was present in these cultures. This is easy to trace if one carefully studies the ancient, archaic cultural layers. However, the source was kept only in one culture—the Chinese. This is not surprising because since time immemorial the Chinese have a tradition that orders to not destroy any books, any sources of information.

According to the story mentioned above, after obtaining a written fixation, this source of knowledge got the name 山海經 –Shan Hai Jing (Catalog of Mountains and Seas, or Book of Mountains and Seas).

Since "God" the Creator created a Human, and not just a separate race and nationality, the "Catalog of Human Souls" was left as one "Instruction to a Human" for all humanity; regardless of what nationality a person belongs to—Chinese, American, Russian, German, Spanish or Chukchi. Therefore, it does not matter that this book is preserved only in Chinese. The important thing is that it is preserved. Especially since the Chinese language, as it is known, is the most ancient language, preserved in civilization. And, it is a very rich language. Due to the richness of this language, images-etalons, which carry information about the structure of human psyche, can be understood and translated into descriptions of individual *Homo sapiens*, as well as recipes, which are practically applicable in their real lives. Probably due to this reason some researchers call hieroglyphics "the language of gods." As our research practice has shown, hieroglyphs really do contain deep meanings of objects and concepts fixed in a language.

PART 4

WHAT IS HUMAN PSYCHE ("SOUL") AND WHAT IS ITS STRUCTURE?

Despite the fact that modern science still tries to argue that humanity, this planet and the entire Solar System were not created by "God," scientists have already made an observation that "God" created every living and nonliving thing on this earth with an innate life "orbit," or, in other words, with a program.

Everyone and everything on this earth, including the planet itself, has a clear program of life and actions. It has been proved long ago that if Earth or any other planet in the Solar System deviates from its orbit by just a few degrees—a catastrophe will occur. However, science is silent about the fact that *Homo sapiens* is an exactly the same hard-coded system; probably because prior to discovery of the Catalog of human population no evidence to support this existed. Although using simple logic, an "empty" born creature (free from a program) could not have possibly appeared in a system that is arranged this way since that would threaten to disturb the common system and create disasters.

It is even more difficult to imagine that "God" could have created a "blank" Human, the "Crown of His Creation," the most complexly created creature on this planet. In this regard, views of some philosophers and scientists who believe that human children are "tabula rasa" ("a blank slate"), on which supposedly anyone can write anything, seem very ridiculous and absurd. The emergence and existence of such creatures in the natural system would be nonsense, and it is very difficult to suspect our Creator of miscalculation and unprofessionalism. And, the ancient treatise with descriptions of human programs found by Andrey Davydov fully confirms this.

However, some philosophers and especially mystics have little doubt that every person is programmed. Although in their language, the word 'program' is replaced with 'fate', 'karma', 'destiny', 'predetermination', and so on. However, first of all, definitions of each of these names are not the same as the definition of a 'natural program', and, secondly, from our point of view, it looks a bit archaic. Opting for the language of the computer age, we prefer to use the word 'program'.

Especially after the discovery of the "Catalog of Human Souls," the fact that a human and a computer have a similarity became obvious: neither one nor the other works without software. The "soul" of a human is, using the computer language, his "operating system." A computer without software is just a bunch of metal, and at best can be used as a cutting board or a shelf. And, a human without an "operating system" (in other words—"Soul") is just a physical shell that cannot be called a human, as it cannot function as a human. Compared to this "exemplar," even mythical zombies are more functional, as they have an inbuilt program to perform certain actions.

*****

So, it was found that people as well as other living and non-living natural objects on planet Earth are born with a particular program on the inside. A 'program' ('an operating system') is the psyche, the "soul"; the very "soul" that "God," according to the story, breathed into a human body during his "Act of Creation." From our point of view, words 'mind', 'soul', and 'individual subtype program' (although there is a purely scientific name for a 'program'—an individual archetypal pattern) are synonyms, and all of these names do not change the essence of the phenomenon.

However, the fact that a human is a bio-robot in any case does not equate him with animals, which are also born with a program. A human as a biological being has categorical differences from animals. "God" gave opportunities to think, to learn, to improve, and, therefore, to transform, to change, to evolve only to humans. And, the Creator also awarded humans an ability choose, as well as creative abilities.

Through all of these processes, which cardinally differentiate a human from any animal, the Creator gave a human being an ability to change his natural program, or rather to replace it with one that is more complex, more perfect, has greater abilities. No other object of nature has similar potential. "The Lord" thus emphasized a human being among other natural objects, and set him above them all: "…and let them have dominion over the fish of the sea and over the birds of the air and over the cattle and over all the earth and over every creeping thing that creeps on the earth…"

However, the Creator of a Human Being for some reason has not awarded humans with neither an innate knowledge of their Program ("Soul"), nor following of this program "on automatic," as it is the case with animals. We are not attempting to judge the reasoning of the Creator, but it is likely that by not giving people an innate knowledge of themselves, "God" wanted to stimulate them to cognitive processes. According to information received from the ancient source of information, "God" expects learning, awareness, studying from a human because Spiritual and physical transformation, Spiritual and physical self-improvement, that what is called "Spiritual Development", "Spiritual Growth" are possible only through these processes. (As it turned out, physical transformation without Spiritual

What Is Human Psyche ("Soul") And What Is Its Structure? | 35

transformation is not possible.) That is why, from our point of view, a burning desire to know himself, and, primarily, a desire to know his "Soul" is present in every human, it is obvious. After all, from a very young age a person wonders: "Who am I?", "What am I?", "What is my purpose?", "What is my mission?" However, the thing is that the Book, from which this could be found out, has disappeared from this civilization due to an unknown reason for many centuries.

We suppose that the Creator of humanity has left the "Catalog of Human Souls" to people with some, let's say, parting words: "Here you are, human, here is an instruction to yourself—study and follow the Path that I defined for you." Most likely, this was done so that every human did not rush about his whole life in search of himself, but instead knew himself, and spent his time on self-development, self-improvement; self-improvement in terms of achievement of compliance of his own self, as an individual, with the "Will of the Creator." Without this "Instruction" people are forced to spend a significant part of their life in search of their selves, and the remaining time is spent on excogitation of their selves (more on this will be in the part on artificial images). As a result, life passes, but without self-improvement, without transformation; people simply do not have the time or the energy for the processes of self-development. And, people do not know what to develop since they do not know themselves.

Also, from our point of view, our Creator left the "Catalog of Human Souls" as a kind of an "Instruction to humankind," so that a human does not get too fond of fantasies about his own self, and carries out the will of the Creator of what he should be. The Creator's worry in this regard is understandable since *Homo sapiens* is the only object in nature, which has a psychological function called "imagination" (more on this below).

However, despite the fact that the "Catalog of Human Souls" was carefully left by the Creator to humanity, and was once present everywhere, in all cultures of this civilization, the Creator still insured the "Crown of his Creation." Namely, made sure that no matter if a person knows his program or not, he still functions according to this program, lives on its basis regardless of whether he is aware of this fact or not. It is for this reason that the descriptions of people from the "Catalog of Human Souls" are almost always a 100% match with the characteristic qualities, algorithms, functioning of real people—they are recognizable.

*****

The whole question is, to what extent and degree do people realize themselves as "God" created them. According to our data, if a person does not receive information about his program from the "Catalog of Human Souls," the percentage of conformity to the potential given to him by "God" is extremely small—people "bury" about 95 percent (if not more) of their talents "in the ground."

Also, it was discovered that if for some reason a person did not get his program from the "Catalog of Human Souls," then he begins to fantasize about his own self: who he is, how he functions, what his goals and objectives are, etc. In essence, he attempts to "create" the program himself. However, he is always unsuccessful. Even worse: he thereby causes great harm, enormous damage to his psychophysiology. How and why this happens from a scientific point of view, we will also discuss in greater detail in the part on artificial images. However, in terms of religion, this occurs because a person deviates from the "Path" that "God" intended for him.

Deviation from the "Path" that "God" intended is punishable. And, one should not hope that the Creator is too busy and will not notice this. Yes, most likely our Creator really has many other things to do instead of watching His every Creation. Therefore, as we have found out, conformity of the real life of a human (that is, the program implanted by "God") is being monitored through natural mechanisms inside the human himself. The Creator organized this process so that all human fantasies on "Who am I?", "What am I?", "Why do I live?" work like a virus in a computer, causing great harm to his organism. Automatically. It is not necessary for "God" to watch each person day and night, He created this system in the beginning, when He created a Human as part of his program. And it works. By leaving the "Catalog of Human Souls" to humans, the Creator wanted to protect His children from harm that they might cause to their own selves.

*****

Therefore, the "Soul," according to the ancient source of knowledge, which was discovered by Andrey Davydov, is that Program which the "Lord," the Creator breathed into a human at the "Act of his Creation." "Soul" is the inner "filling" of a human. "Soul" is the basis of all that what is called by the word 'human'. A body is a receptacle in which "Soul" lives and functions. Therefore, from our point of view, with the discovery of the "Catalog of Human Souls," the still ongoing debates in the field of psychology on what is primary, the psyche ("soul") or the body—no longer have the right to exist. These disputes are an indication that psychology is a science that has no clue about the main subject of its research. As it turned out, "Soul" is primary since it is the hidden inner core of any processes that occur to a human being, the foundation of his life, functioning, manifestations of qualities, talents, and so on.

However, contrary to views that a person's body is just a "mortal shell"— neither a body without a "soul," nor a "soul" without a body can exist separately. Body and "soul" are one whole. Therefore, considering a person holistically, rather than as a "dismemberment" as it is done by some researchers, we use the word 'psychophysiology', implying that body and psyche ("soul") are an organic whole.

*****

Psyche ("Soul," Program) of a human can be compared to part of an iceberg that is underwater. However, it is psyche that commands all processes of human life. Also, note that psyche is not intellect, as it is commonly believed among modern psychologists. Human intellect, one function of which is imagination, can dream as much as it wants: make plans, come up with scenarios of actions or self-presentations. However, that what is hidden even from the carrier himself—his psyche, his Program, his "Soul"—actually dictates him how to live.

*****

We would like to expose another myth. Today, psyche is considered unknowable because it is supposedly invisible, inaudible, intangible, etc. Doubts that the "Soul" exists are based solely on that supposedly it cannot be recognized by any existing human senses. That is not true because the "Soul," all of its qualities, are directly expressed in manifestations of a person, and one can see, hear, smell, touch and taste them. Any person looks a certain way, smells a certain way, sounds a certain way, moves, speaks, makes some actions, in other words, expresses himself. And, he expresses himself differently from other people (hopefully no one will argue this with). Therefore, to formulate more correctly: "Soul" is invisible to those who have problems with vision, "Soul" is inaudible to those who have problems with hearing, and so on. However, first and foremost, the "Soul" as a mechanism is invisible to those, who do not have information from the "Catalog of Human Souls" about this or that person. Qualities of individual "Soul" are always directly reflected in external manifestations of any individual, it is only necessary to be able to recognize them.

*****

"Soul" (Program) manifests itself in a person through six factors: intellectual, physical, nutritional, emotional, sexual and environmental. This six-factor breakdown used for descriptions of psychophysiological manifestations of *Homo sapiens* was also created and put into research practice by Andrey Davydov.

Prior to this, no one applied such a complete and detailed method of describing a human: neither in medicine, nor in, and especially so, in psychology, where things are in general not good with specifics. In psychology, no one has ever done descriptions of an individual on six factors until Andrey Davydov, the author of the discovery of the Catalog of human population ("Catalog of Human Souls"). Even though animals are described in encyclopedias much fuller. Description of each subspecies of animals typically contain the following information: appearance, physical characteristics, sex life, abilities, what, how and when to eat, what its attitudes towards the same subspecies and members of other species are, as well as climate, living conditions, etc. Sometimes information even about the emotional sphere of a described animal is presented, as in, for example, works of the famous German scientist-zoologist Alfred Brehm. (Although from our point of view, those are very subjective judgments). The only factor, detailed descriptions of which you will not find in an encyclopedia on animals is the intellectual factor. At maximum, there might be a mention of, for example, a possibility of training or an ability to perform multi-step tasks (as in animal psychology, for example), classified as extraordinary cleverness of any particular animal or an entire subspecies, such as ravens. However, there is no serious discussion about animals' ability to think or to imagine. Their intellect is completely replaced by instincts and reflexes, and that is enough for survival of animals. As far as the emotional sphere of animals, in comparative analysis with a human it functions very primitively. However, despite the primitivism of animals compared to a human, their descriptions are more complete and informative than information about people, which is usually limited to standard form data like gender, age, race, nationality, education, and profession; or it is a purely medical description that is one-sided, and considers only functioning of physiology.

Compared to humans, animals generally have a much more primitive structure and their potential is even more primitive. Potentially, the difference between the work of factors of *Homo sapiens* and animals is so categorical that it is impossible to discuss any kind of analogy. For this reason, it is very amusing to hear suggestions that humans evolved from apes. With knowledge about functioning of human psyche and the Catalog of human population, this sounds as funny as if someone stated that one of the Formula One cars suddenly "self-assembled" at a junkyard from a pile of trash. Although, on the other hand, if human potential does not open in its entirety, then a human is not very different from an animal. However, that does not change the fundamental differences, or the number of factors. In case of a human there are six factors, and the whole question is how each one gets developed, but that is a different matter.

Few specialists directly connect human psyche with the work of all these six factors. In vain. It is psyche ("Soul" or Program, call it whatever you like) that is the basis of work of human intellect, physiology, nutrition, emotional and sexual factors, as well as the environmental factor. It is the natural Program that sets the physical parameters of a person, determines his physique, the work of his internal organs and systems; how, when, in what algorithms his intellect will work, what specific activities develop him, etc.; what, how and when a person will eat, etc.; what emotions, how and when he will express, etc.; what his sexual potential, needs and preferences are, etc.; what the requirements as far as climate, housing, interior, communications and travel for that particular person are, in what areas of professional activity he can realize himself, how he will make money, how much, in what ways, how he will interact with others, on what basis he will communicate, how he will present himself, etc.; and much, much more. Behind all of this is the "Soul," or the program.

PART 5

THE "CATALOG OF HUMAN SOULS" EXISTS. INDIVIDUAL SUBTYPE PROGRAMS AND MANIPULATION MODES OF *HOMO SAPIENS*.

We apologize for the abundance of scientific language in this section, but it is most suitable for a correct explanation of what the main object of our research (the Catalog of human population) is.

Thus, as it turned out, the "Soul" of every living creature on planet Earth (including *Homo sapiens*) is a program. The discovered ancient book describes structures of human psyche ("Soul") as 293 unique models. Therefore, the source of knowledge of what the "soul" of a human is and what is its structure was named the "Catalog of Human Souls" by its discoverer Andrey Davydov. However, there is also a scientific name of this source—the Catalog of human population. This does not change the essence of this Book: 'a catalog', 'an encyclopedia', 'an atlas' are synonymous in this case. The Catalog of human population is an encyclopedia of *Homo sapiens*, from which it is possible to obtain detailed descriptions of people. Including those people whom neither we, the researchers of the Catalog, nor the requestor of information have never met or seen in our lives and know nothing about.

*****

In contrast to encyclopedias on animals, where they are described by purely external features, in the Catalog of human population people are described based on qualities of their "Soul." Or, to put it in scientific terms, based on qualities and characteristics of their individual subtype program. Although as we mentioned above, the qualities and characteristics of this program are directly expressed in the appearance of an individual, in particularities of structure and operation of his physiology, in his nutritional, emotional, and sexual preferences and algorithms of functioning, in the way that his intellect works, as well as in everything that has to do with his environment (climate, housing, interior, education and occupation, relationships with others, etc.). "Soul" (an individual subtype program) is the basis of functioning of all of these factors.

Individual subtype program is the foundation upon which the whole human psychophysiology works.

Also, it is the individual subtype program that distinguishes one person from another. Although in the case of representatives of *Homo sapiens*, without a catalog it is impossible to clearly (as in the case of animals) see the differences between a representative of one subtype and another because external and internal physiological structure in humans is the same: two hands, two legs, one heart, two lungs, two kidneys and so on, no wings, horns, fur, scales, tail, beak, and so on. This external similarity distinguishes *Homo sapiens* from representatives of other biological types. However, it also misleads one to assume that all members of the species of *Homo sapiens* are the same; although everyone knows that this is not so, and there are many confirmations.

*****

Discovery of the Catalog of human population quite distinctly marked a new approach to a human: humankind is a biological type within which there are about 300 subtypes. With the discovery of solution to the riddle of this ancient Book, it became possible to talk about the characteristic patterns of functional range of human form in different time cycles. This provides an opportunity to separate the biological type *Homo sapiens* into certain groups (groups, which are carriers of certain characteristics, patterns).

We, as researchers of the Book of Mountains and Seas (Shan Hai Jing), considered it logical to name groups within the biological type *Homo sapiens*—subtype (subspecies) groups or subtype structures. Since the term 'subspecies' is a taxonomic unit, located a rank below in the systematization of plants and animals, but represents totality of separate populations of the same species, members of which differ by specific and rather stable characteristics from members of other populations of the same species.

Subtypes (subspecies) of *Homo sapiens*, as it turned out, differ from each other in a unique complex that consists of an individual subtype program and three modes of correction to this program: the suppression mode, the balance mode and the stimulation mode. We named correction modes of a subtype program—manipulation modes. Manipulation modes, as well as an individual program, are unique in the case of each subtype. On the basis of categorical differences between combinations of an individual subtype program with manipulation modes, each subtype is endemic; that is— different from all the rest.

*****

The study of Shan Hai Jing revealed that every person, belonging to some subtype, has stable characteristic properties of this subtype; and, regardless of race, nationality, place of birth and residence, parental guidance, and so on, as these are only minor correctors, which do not change a subtype program. Though, the same pattern is observed with animals. Living in different areas (as some animal subspecies, as well as human subspecies can be found almost all over the globe), the representatives of the same subtype may have different adaptive properties, but these properties do not in any way alter the subtype program: a horse remains a horse, a bear remains a bear, a bull does not turn into a ram, and a snake does not become a crocodile. Only here it will be a pony, somewhere else a percheron, and elsewhere—an Arabian horse. Regardless of the territory where an animal lives, it retains stable properties of its subtype. We see a similar situation with humans: the representatives of the same human subtype might have different skin color, be of different nationalities, but each retains those personal qualities, life algorithms, preferences, talents, functioning that are inherent in his subtype.

Cultural factors, national traditions, even parental upbringing (which, by the way, plays a big role for representatives of some subtypes) are only additions, layers, and do not alter the subtype program of a human. Since, in the case of biological species *Homo sapiens*, children are not direct continuation of their parents, as it is the case with animals. More often than not, children and parents are members of different subtypes, and that is the basis for existing problems of parents and children in this civilization. If one does not know about the Catalog of human population, then this problem is practically unsolvable. However, knowledge of this information and knowledge of exactly which programs parents have, and which program a child has, fully eliminates the problem of "Parents and Children."

Certainly, as with any other natural object, the role of the environment is enormous in the life of any human being. If a seed gets in the wrong soil, it might never sprout. This principle works with humans. Within certain limits, the environment shapes a person through engaging various program qualities that he has since birth. However, influence of the environment is limited. And, neither the environment, nor parents, or society are able to change a program given to a human by "God." Environment can favor an individual in the sense of creating suitable conditions for his life, helping discover and develop some of his skills and talents, but it can also severely deform, break, and even destroy him. However, no environment can change the "Divine Will" in regard to a human, "Will" that is implanted as qualities, characteristics, functions, goals and objectives of his individual subtype program. A long time ago the Creator has predetermined what each one of us must be, and how each one of us must live.

*****

"God," the Creator of humanity, clearly predestined what each person must be, how he must live and what he must do. This program is "wired" into each person on the inside. And, it is the subtype program, the "Soul" that distinguishes one person from another. Variety of qualities and functioning of human subtypes is huge. This is not surprising because all nature created by "God" on this planet is very diverse. For example, birds. Despite the fact that all birds are members of a single biological type, there are many subspecies, and representatives of these subtypes are very different from each other: there are huge birds (an ostrich), and there are small ones (a hummingbird). There are predators that eat only meat, and there are those that eat mostly plant foods. There are waterfowl, and there are those that do not swim. There are those that spend most of their lives in flight, and there are those that have wings, but practically do not know how to fly. So, on the one hand, they are all different, with different qualities and different functionalities, but on the other hand, they are all birds, not horses, or crocodiles, or tigers. Exactly the same diverseness was created by the Creator in regard to all other natural types; both living and not living natural objects: trees, plants, animals, minerals, stones and so on. And, as it turned out with the discovery of the "Catalog of Human Souls"—it is the same in regard to humans.

The inner core of a human, his psyche, "Soul," or, using the religious language—the "Divine Program," makes the physiology of representatives of various subtypes of type *Homo s.* work differently. This is by far the main problem in medicine: it cannot find uniform methods of treatment for all people, one type of medication for all, and so on. Here are some of the most simple, easily observable examples. Individual differences in work of physiology allow some people to, for example, walk around practically naked in the cold, while others get sick even from a small draft. For some, it is not a problem to eat only vegetables, while for others lack of meat, fish, protein foods turns into very serious health problems.

Some people can think, calculate the variants and make decisions quickly, while others make decisions only after a lot of thinking. Some have a good ear for music, and others do not, but are able to run long distance without stopping and reach the finish line first. Some are polyglots, and for some even their native language is difficult, for example, in terms of grammar, spelling, but they are genius in mathematics. To some getting up early in the morning is not a problem, and they do it even in the absence of necessity, while others cannot put themselves to bed early, and "burn the midnight oil" because they can function more productively at night, but cannot rise at 6 AM and work well during the daytime. These are very simple examples that are familiar to everyone. And, even though people are used to explaining this using reasons that have nothing to do with work of their psychophysiology, their psyche, their "Soul"—"God" created a human in such a way that everything in his life depends on the Program.

A Program ("Soul"), or to put it in scientific terms—an individual archetypal pattern, fully defines all motivational mechanisms of human behavior, his functioning, abilities, preferences, goals and objectives. And, as it was stated, they are completely different for representatives of different subtypes of *Homo sapiens*; and this is the basis of individual differences between people, as individuals with a specific sets of behavioral qualities.

*****

Individual subtype program ("Soul") begins to work at the moment of birth. With life, "God" as if breathes into a tiny creature all those qualities and functions, which he would like to see in him; all that what the Creator expects this or that person to realize.

"Hour X" as time of birth of a human, starts up his subtype program. The "Soul" begins to work, and a human—to live. And, he will live while his "Soul" is alive in him; in other words, while the subtype program operates in his body as a mechanism, and on all six factors. If the work of one of these factors stops, then breakage of other factors will follow: fast, as in the case of, for example, physical and nutritional factors (starvation, disease), or slow, as, for example, in the case of sexual, emotional or intellectual factors.

It does not matter how a person is born, whether in a natural way or, for example, with the use of cesarean section. Individual subtype program starts to work at the moment of cutting of the umbilical cord. It does not matter when a person was conceived, how long the pregnancy lasted, and so on. The only important moment is the time of detachment of his body from the body of his mother in the form of cutting of the umbilical cord. This, not something else, is considered the moment of birth.

In the course of study of the ancient source of knowledge about humans (Shan Hai Jing), it became clear that an individual subtype program ("soul") of *Homo sapiens* gets implanted strictly in accordance with natural phenological cycles. Therefore, in order to determine to which subtype this or that person belongs, and to find his description in the "Catalog of Human Souls," it is necessary to know the date of his birth—day, month and year. However, the requirement to know the exact day, month and year of birth in order to get information about a person from the Catalog of human population has to do specifically with phenological cycles, rather than astronomy and especially not astrology, or numerology, or something related. (Note that the year of birth does not matter, but what makes a difference is whether it was a leap year or a common year.). For those who are not familiar with the science called phenology—in a nutshell, it is a system of knowledge and sets of information about seasonal phenomena of nature, time periods of occurrences and reasons that define these periods, as well as the science of laws of cyclical changes of natural objects and their complexes, associated with annual movement of the Earth around the Sun.

In nature, all bio-forms (plants, animals, birds, fish, insects, reptiles, etc.) appear during their time period. Scientists have recorded this fact a long time ago. At the beginning of the last century, a German zoologist, Alfred Brehm, has demonstrated that each type of living organism appears on earth at a specific time. For example, rabbits are born in late April-early May, deer—in May-June, seals—in March, and so on. "...And the stork in the heaven knoweth the appointed times, and the turtle, and the swallow and the crane observe the time of their coming..." (Jeremiah 8:7) "God's" plan in this regard is more than clear: a human, as part of the earth's ecosystem, as part of nature must submit to the same laws as all other natural objects; because if at certain specific time periods creatures, which we call "a cow", "an eagle" or "a hippopotamus" are born, then exactly the same rule applies to humans.

Right in its sections titles, the Catalog of Mountains and Seas already has an indication of the spatial reference points. The ancient Chinese text begins with the Catalog of Southern Mountains, followed by all other sections of the Catalog of Mountains named according to cardinal directions: Catalog of Western Mountains, Northern and Eastern. Next, there is the Catalog of Central Mountains. It "stands" separately, just like the Catalog of Lands Within the Seas that crowns the whole text. Then, there are: the Catalog of Regions Beyond the Seas of South, West, North and East; the Catalog of the Great Deserts of the East, South, West, North, and finally the Catalog of (Lands) Within the Seas. Catalogs of Mountains are divided. The Catalog of Southern Mountains and the Catalog of Northern Mountains are divided into three parts, and the Catalog of Eastern Mountains and the Catalog of Western Mountains into four parts. This reminds of phenological approach, where division into seasons and sub-seasons are analogous to construction of the Catalog of Mountains; Catalog of Central Mountains is divided into twelve parts.

The birth of a particular subtype structure is strictly tied to the date of birth. However, in the case of research of Shan Hai Jing, things are not as simple as, for example, in astrology. At least because, as you know, there are 365 days in a year, or 366 if it is a leap year, while there are less than 300 subtype structures of *Homo sapiens* described in the Book of Mountains and Seas. To be more precise, there are 293. However, Andrey Davydov has solved this problem long time ago, and finding any description is no longer an issue. It was also discovered that in most cases, if a person was born during one of the leap years (no matter which one), then his date of birth as if moves one day ahead. For example, people who were born on March 1st of leap years have the same subtype program as people who were born on March 2nd of common years. However, this rule does not always apply. In this sense, in addition to a year, month of birth has a meaning. As for the time of birth within one day—this factor is not needed in order to find descriptions of this or that person in the Catalog of human population, unless there is data that this person was born close to midnight (and, in this case we also have the technology to determine where the description of this person is located in the Catalog of Mountains and Seas).

Many of those who learn about the Catalog are interested in how 293 individual subtype programs are distributed if there are 365 or 366 (leap years) days in a year. We found that some subtype programs get reproduced a few days in a row. However, this does not mean that people born in this period are absolutely the same because manipulative modes to these individual programs are often completely different. So, in simple terms, these people are not like each other 100%, but they do share the same subtype program, individual qualities and algorithms.

*****

Another important fact is that each subtype program consists of two parts. One of them, the author of the discovery of the "Catalog of Human Souls," Andrey Davydov named "Active," and the second "Passive."

The "Active" part of a program of a person is responsible for the period of human activity. It is a period of real action, but also a period of active energy and strength expenditure. Therefore, every person requires rest from time to time, and not only in the form of sleep, but also in the form of some kind of activity. Not for nothing there exists a Russian saying: "The best rest is change of activity." Human activities from the "Passive" part of his individual program do not take strength, but rather help a person accumulate it, turn on regeneration processes. "Active" and "passive," to put it very raspingly and primitively, are like two different human characters because each part of the human program is recorded by different images that give a different character, style of actions of a human, and algorithms of his life. In practice, this manifests in the so-called duality of any person. For example: Mr. N is quiet and harmless at work, but at home he is active, loud, and despotic; a different Mr. N uses his intellect at work as an analyst, and during the time of rest chooses physical exercise and, as they say, "does not use his head at all." Some other Mr. N behaves as a leader in communication with several people, but when he is alone with someone—he completely subjects to another's will; and so on and so forth. There are many other examples, but the main thing here is that every person (a man or a woman) is "dual" by nature, double-natured. This "duality" is necessary because one cannot be passively recovering while at the same time actively spending himself.

Without going into a detailed explanation of what these are, we will just say that, according to our observations, in some cases, twins can share a single subtype program; meaning that one lives according to the "active" part of the program, and the other based on the "passive." We do not know yet why this happens, more research needs to be done on this, but this is the reason why, despite the fact that both twins belong to the same subtype, they differ in their personality traits, characters, algorithms and so on. According to our research, this does not always happen in the case of twins, but such cases have been observed.

*****

Thus, with the discovery of the Catalog of human population it became possible to determine to which subtype this or that person belongs, and get a detailed description about him or her, using minimum knowledge (only the date of birth). This is a fundamentally new approach to the study of *Homo sapiens*.

In parallel, the science of psychology gets its subject of research. It was substituted by anything during all the years of existence of this science: study of external human behavior, sphere of the consciousness/subconsciousness, study of personal experiences, human intellect, and so on. However, research of the most important—the psyche ("soul")—has been forgotten. Although the term "psychology" originated from Ancient Greek ψυχή—"soul;" λόγος—"knowledge." Therefore, psychologists must study and know the "soul." And, know that "soul" is not intellect; after all, intellect is not psyche, it is only a part of psyche. Intellect is only one of the six factors, and it is far from being the most important one. If, for example, the physical or the nutritional factor (starvation, disease) stops working, then work of intellect can be forgotten.

*****

It should also be noted that the Book of Mountains and Seas contains descriptions of natural mechanism, which allow making corrections to psychophysiological parameters and behavior of *Homo sapiens*.

Apparently, our Creator took care of the "Crown of his Creation" in this sense as well, and gave each human correction modes of his psychophysiology as a tool for self-control.

However, it turned out that using that same natural mechanism that is built into every human, people are able to influence each other with great effectiveness. Andrey Davydov, the author of the discovery of the Catalog of human population, named natural mechanisms of regulation and self-regulation of a human—manipulation modes.

Each person has three manipulation modes: suppression, balance and stimulation. Each one of these modes has a specific effect on the whole psychophysiology of a human.

For example, if we consider the suppression mode, as a mode of self-regulation of a system called "human"—this mode rather powerfully relaxes psyche and body of an individual, brings peace and joy, pleasure, and satisfaction. The balance mode is named "balance" because it balances, stabilizes psychophysiology, and leads to a harmonious, comfortable state. The stimulation mode is a strong irritant and a powerful motivation for an individual. This, of course, was a very brief description of manipulation modes.

In regard to application of these same modes to a person (not by himself, but by someone from the outside), then, for example, during transmission of suppression mode in his direction, a person experiences strong relaxation and often a state similar to reaction to narcotic drugs—he is lost. He is disoriented in the environment; analysis of the situation and the individual who transmits this mode becomes impossible. Therefore, this mode of self-regulation, if it is translated from the outside, is a mode of submission to the transmitter. In the case of transmission of the balance mode, the receiver feels very comfortable, gets a feeling that the transmitter is his good friend, who shares his thoughts, preferences, goals and objectives. As for transmission of stimulation mode, the transmitter of this mode is simultaneously perceived with irritation and indignation, but on the other hand stimulates the desire to patronize him. Also, through transmission of this mode one person can motivate another person to any kind of action.

It does not matter which race, nationality or gender is a person to whom manipulation modes are applied. In this sense, representatives of the entire human population are made in the same way. Each individual has an individual subtype program implanted by nature, as well as individual modes of self-control/management of life processes, and behavior in particular.

Our studies have shown that human manipulation modes are a subtype's peculiar language. If a person is not spoken to in this "language," then he or she practically does not hear the interlocutor, does not want to understand, does not accept or simply rejects. Our research of Shan Hai Jing has shown a tight link between this "subtype language" and individual subtype program (human "Soul"). Therefore, manipulation modes is the only language that "Soul" hears and understands. From our perspective, manipulation modes of a human were created by the Creator as the only way of effective, constructive, non-deforming, and secure communication.

Based on the fact that application of natural manipulation modes in communication with any person always exceptionally favorably affects his psychophysiology (and not on the basis of the meanings of what manipulation is usually considered) these modes were named manipulation modes. Contrary to the belief popular both in psychology and in everyday life that manipulation is a process with a negative connotation, we consider manipulation based on other meanings from dictionaries. We consider manipulation as any purposeful action with a subject (Philosophical Dictionary).

From out point of view, communicative impact that leads to actualization of quite specific motivational states in a subject, and makes him behave in a certain way, should not necessarily be beneficial to the manipulator and to the subject of manipulation. From our point of view, this depends on ethical and moral qualities of a person who is applying manipulation modes towards another person. Just like a knife cannot be regarded only as a murder weapon. A knife as a tool can also be used as a cooking utensil, as a scalpel in the hands of a surgeon, and so forth. From our point of view, it all depends on what kind of person uses this or that tool, and with what intent.

Of course, being very powerful, this purely natural mechanism can be used by some to have the other people submit, in order to use them for some purposes. Indeed, when personal manipulation modes are applied to a human, he or she forms a complete willingness to do anything for the transmitter; even that what he or she would never do without the use of this mode. Moreover, everything that this person was motivated to, he explains to himself and others as solely his decision, a manifestation of his free will, and an absolutely voluntary action on his part. And, indeed, this is the case because motivational mechanisms get triggered on the inside. This is very different from a situation where a person is forced to do something, and he realizes that he has to do it, that he was forced to do it, and lacks an internal desire to do it. However, we consider use of manipulation modes with immoral goals as unacceptable, inhumane, immoral, and contrary to the "Creator's Intent."

However, as shown by our studies, the dream of obtaining such means (means of total control over another person) does not leave humanity. For many centuries and even millenniums, scientists, motivated and funded by top leaders, have been trying to find a method exactly like this. At all times the value of this information has been so high that any methods were used to obtain it: genocide, war, inquisitions, sects, concentration camps with experiments a la Dr. Mengele, and so on. In order to get hold of this method, people were used cruelly: burned, drowned, cut up alive, dismembered, crucified, frozen, put in the terrible psychological conditions, and so on. However, despite the fact that countless human lives were laid on the altar, a unified theory that would offer individual models of influence (and with a perfect result) was not found. To date, attempts of civilization to control a human being remind of attempts to study a living cell, which, as it is known, is not visible without a microscope and simply dies under a microscope.

Alas, modern "Mengeles" continue to mangle the human body in search of management "buttons," and marketers continue to obsessively offer yet another pizza. Such tactics, from our point of view, indicate lack of knowledge on how and based on what the human mind works. Therefore, talks about the possibilities of manipulating a human often cause not only fear, but also skepticism: "It is definitely impossible to manipulate me." Although as we have found out that is far from the truth.

Manipulation modes are part of structure of the human psyche, and they are implanted at birth on the level of reflexes. No human being can resist their influence. With this tool, anything can be done with a human resource on any scale. However, as it was already mentioned, this tool puts enormous responsibility on a person who uses it, and first of all, before the Creator.

Whether one likes the existence of natural mechanism of any person called "manipulation modes" or not, it exists anyway. And, what is no less important, this is part of the "Creator's Plan." Therefore, in the absence of the Catalog of human population it is, of course, possible to continue to deal with *Homo sapiens* as with animals, while discussing humanism, goodness and morality and other things unrelated to this process. This is widely practiced in civilization. Consider a method that in the case of animals is called taming, and in the case of a person for some reason is referred to by different words: upbringing, education, re-education, etc. (Zoopsychologist are very well familiar with this method as the "method of positive and negative reinforcement.") However, unlike subtype programs of animals, in which only instincts and primitive reflexes are recorded, human programs consist of much more than that. From our point of view, Pavlov's (a Russian researcher, who was involved in elaboration of certain reflexes in dogs) method that has become popular worldwide should not be practically the only available method of influence on a human being in the XXI century. A Human was created by "God," and not in the image and likeness of an animal, but in His "Image and Likeness." Therefore, communication with each person must be based on more complex principles than induction of salivation, like "Pavlov's dog." It must be more variative and gentle than with animals. In their pursuit of power, people should not forget about moral basis and principles.

*****

In conclusion to the presented information on human manipulation modes, another very interesting fact is worth mentioning. While studying the Catalog of Mountains and Seas (Shan Hai Jing), Andrey Davydov found out that each one of the three manipulation modes of a person is some subtype's individual program. Therefore, it turns out that people within the same subtype are not just blood relatives (blood because we suspect that blood may be related to the "operating system" of a human), but it is also possible to say that they have very close kinship ties with three subtypes (meaning, carriers of suppression, balance and stimulation modes). Since they are in a natural interrelation through manipulation modes.

However, we will not go into the description of this rather complicated natural mechanism. We will just say that loneliness of a person in this world is a 100% artificial phenomenon. In nature everything is arranged in a way that there is no loneliness; there is a community of autonomous individuals who are in constant interrelation with each other. Only, *Homo s.* in this civilization really is lonely. And, it just seems to him that he is lonely and alone only occasionally. In fact, he or she is always alone. And, no matter how many people are near because there is no real contact with them, there is only fiction, an illusion of contact. No one really understands him or her, and cannot understand, even if one really wants to because that person does not even understand himself. The reason for this is lack of information about oneself and other people from the "Catalog of Human Souls." When there is no clear knowledge of who is in front of you and what to expect from this person, then there is nothing but mistrust, fear, confusion, and sometimes protest and anger.

However, as our studies have shown, things were made quite differently. Our Creator has created people in such a way that no one ever has to be lonely. Therefore, a person who knows about the "Catalog of Human Souls," is familiar with his program—knows about the fact that there are about 20 million people on earth, who by nature are closer to him than his relatives. (To be more specific, today it is over 23 million, if we divide population of the earth—7 billion people—by 293 subtype structures.) Since these people fully share all of his preferences, thoughts, desires, goals, and so on. We are referring to members of one human subtype. And, a person who knows his program knows all of these 23 million people very well. And, he knows what are the criteria to find them, and where. However, most importantly, he knows how to communicate with them in order to get pleasure, comfort, and use from communication, instead of irritation, tiredness or damage.

Any person who uses information from the "Catalog of Human Souls," and knows not only his own manipulation modes, but also the date of birth of people who are natural carriers of these modes—becomes the owner of widest opportunities for communication. This person can easily find an ideal partner for any purpose, including happy family life, and a person whom he will love greatly, and a person who will be madly in love with him, literally adore him; and, of course, true and reliable friends who will never betray. What kind of loneliness can there be in this case? A huge mass of people is at the disposal of a person familiar with the "Catalog of Human Souls." In addition to the 23 million people discussed above, another 70 million people are carriers of his three manipulation modes. Among them, one has no chance not to find those whom he or she is looking for. Guaranteed.

Therefore, each representative of any subtype has a natural connection (through the subtype program and manipulation modes) with more than 90 million people who are close to him in spirit. And, close in a literal sense of the word since they are linked by natural "Soul." It seems to us that from such a colossal number of potential contacts any person can organize quite a large circle of friends, and in his case he will not be alone or lonely. This is another opportunity that a person gets with information from the Catalog of human population; in particular—knowledge of manipulation modes.

*****

Thus, as we uncovered, every living organism on the planet Earth has a fixed program from the moment of birth, and cannot go beyond its possibilities; even a human being. (However, in the case of a human, this rule works only when he does not know his program from the "Catalog of Human Souls," and does not self-develop and improve.) Therefore, everything (to the smallest detail) in the life of any natural object is programmed by the Creator. Therefore, our planet is a mechanism with strict reproduction of programs in a specific sequence for each natural subtype. It is no wonder that some modern researchers compare the functioning of our planet to an enormous computer, and jokingly call "God" the programmer.

PART 6

DISPEL OF THE MYTH ABOUT INCOMPREHENSIBILITY OF A HUMAN AND HIS "SOUL"

For a long time *Homo sapiens* remained one of few biological objects, which were not divided into subtype structures, and did not make it into an encyclopedia. Of course, there were some encyclopedias about a human being, such as anatomical atlases. However, one could not get information about what this or that person is like and how to communicate, interact with him. The discovery made by Andrey Davydov allows study of a human according to his subtype characteristics, based not on the structure of body, but on an individual program that relates to the internal, psychical space.

Consideration of each person as a representative of one of the subtypes allows one to learn any information about him/her using minimal personal data (only the date of his/her birth). This is very convenient, and provides fundamentally new opportunities; both in terms of research in this field, and from the standpoint of practical application of the Catalog of human population in daily life.

Lack of a situation where a person is an unknown, an unknowable, an uncomputable value has huge advantages, including an ability to obtain any kind of individual recipes for each person, and solutions to any problems on all six factors: intellectual, physiological, nutritional, emotional, sexual, and environmental.

As it turned out, in order to make contact with another person and turn it in the required direction, it is necessary to understand exactly who your opponent is: how he or she will act, what and how he or she will say, what he or she will seek, and so on. To know all this, it is no longer necessary to meet this person, to watch him, to try to learn something about him from someone, to provoke some actions, to study his reactions, and so on. There is no longer a need to be in any situation with a person in order to know exactly what to expect from him (what actions, what reactions, what decisions) and what to be cautious of with him. It is enough to simply read the description of a person in the Catalog of human population.

Any information about any person can be obtained from the Catalog of human population; information on all six factors: what he is like intellectually, how and how fast he thinks, in what algorithms, what his interests are, etc.; what, when, where and how he eats, etc.; what his physical features, abilities, and talents are, etc.; what emotional displays are normal for him, how powerful they are, how and when they manifest, etc.; what are his sexual preferences, sexual potential, how it gets expressed, in what forms, with what partners, under which conditions, based on what algorithms, etc.; what areas, what type of property, with what interior, and which conditions he prefers to live in; what are the natural talents of this person, how these talents can be realized, including in the professional sphere (although not necessarily in a professional one); how he communicates with other people, on what basis, what he wants, what methods he uses to achieve this, what people should be cautious with in regard to this person; and much more. All this information is now available from descriptions of individual subtype programs from the "Catalog of Human Souls." Now you can learn everything about any person without ever communicating, speaking, meeting with him and without knowing absolutely anything about him.

As it was already mentioned, in order to get information about a person from the "Catalog of Human Souls"—the exact day, month and year of his or her birth is required. In addition to the birth date of a person it is also necessary to know the gender because, in this sense, subtype programs of *Homo sapiens* are similar to programs of animals, where there are some differences between males and females. Subtype program of *Homo sapiens* is the same for both sexes, but there are some differences in the sexual factor, which distinguish functions and behavioral characteristics of women and men.

Since any individual normally lives among other people, the problem of knowledge and understanding of another person is everyone's urgent concern. However, to date there is no source (except the "Catalog Human Souls") in civilization where a specific John could learn how to communicate with specific Tom or Mary. And, not just communicate, but do it in that exact manner that John needs. And, not just get "whatever" from this communication, but strictly that what he personally needs. In an attempt to achieve this John uses one method after another (designed "one for all" as usual) until all methods are tried out, and something finally works at least to some extent. However, sometimes it is too late—the relationship is ruined, the contact is broken, and goals in regard to the opponent are not achieved.

We think that the root of this problem is fundamental difference of internal characteristics of one person versus another, despite the external similarity, and sometimes even kinship. Since it is impossible to distinguish *Homo s.* subtypes based on appearance, it is much easier to tell a rabbit from a donkey, despite the fact that both have long ears. And, that could hold very unpleasant consequences. In nature, when one interacts with a wild wolf as with a dog, with an eagle as with a dove, and with a poisonous snake as with a worm—the consequences are always fatal. However, it is the same with people. Not knowing the person with whom you come in contact, it is impossible to know what to expect from him or her, and how to interact. Of course, unlike it is with animals, by interacting with another person inappropriately, in the physical sense you will not get bitten, scratched or torn apart. (Though this happens sometimes.) However, for example, in psychological or financial sense: biting, scratching and tearing to pieces is not uncommon among people. Unfortunately, this type of "communication" occurs very, very often. As a result, after communicating with other people a person very often has to "lick his wounds."

However, in this respect the situation has changed dramatically thanks to the "Catalog of Human Souls" discovery made by Andrey Davydov. Now, in order to make any kind of contact it is not necessary to gather information about a person from third parties or from other sources of information, there is no longer a need to use the well-known "scientific method of trial and error," and so on. Contact (business, friendly, intimate, and so on) with any person is now done quite easily using other methods. In order to know what to expect, and what not to expect from a person, all that is needed is information about the individual program and manipulation modes of that person. Application of this kind of information guarantees establishment of any type of required communicative contact and with any required results. Any interaction can easily be modeled using information about another person from the "Catalog of Human Souls." And, as a result, any person can get that what he wished, that what he wants and needs from another person.

The success of use of this technology is not affected by the gender of the subject of interest, or by his/her age, educational level or professional affiliation, financial situation or association with any social group. Information about another person from the "Catalog of Human Souls" allows anyone to have an ability to influence, make adjustments to behavior, plans, and objectives in relation to that person.

Consider any of the problems, for example, inside tandems "husband-wife" ("lovers"), "parent-child" that we can safely state are common in 99.9% out of 100 cases. The root cause of any one of them is the same: lack of knowledge of another person, inability to properly communicate with him or her, and inability to influence him or her. These are, so to speak, the "three pillars" upon which in this civilization all communication between one person and another stands. Only because of these "three pillars" parents at some point lose understanding of behavior of their children and control over their actions, and children are not able to "get through to parents," to explain their needs, aspirations, life stance to them. Also, this is the only reason why at some stage men and women go in different directions and live separately—they are simply unable to neither find a person with whom they would be happy, nor find a common language with a person whom for some reasons they selected for family life or for another type of relationship. And so, they live apart, despite the fact that both women and men, regardless of how old they are, do not want to be alone. They want to feel someone's shoulder close to theirs; they want to be needed, loved, and wanted. And, at any age a person needs to satisfy the basic needs, such as communication, sex, and much more. Although one can declare anything on this account and explain it in any way.

To us, all of the problems listed above seem unworthy of wasting one's breath on, as they are easily solved using information from the Catalog of human population. And, the so-called search for "the other half," and mutual understanding and love between parents and children or between a husband and a wife, and how to find a common language with another person, and how to be liked, and how to affect, and so on.

So-called "ordinary people" can dream as much as they want that if he or she is not involved in something, from their point of view, serious and dangerous, they should not worry about their safety or preservation of their well-being. From our point of view, this is a very big delusion. We live in a civilization where each one of us, figuratively speaking, lives his whole life in the area of active combat operations. Regardless of whether he knows about it or not, notices or does not notice it. Only the scale of the combat operations (and, without quotes) is different for different people: for some it is the process of conquest of economical and political influence over entire regions, and for some it is fighting over, figuratively speaking, a rusty frying pan. It so happened that in this society, in any type of relationships between people, actions against each other that cannot be classified in any other way except by the word 'war' are present: in business areas, so-called "friendship," and even family. And, in reality it is very difficult not to notice this fact. As a rule, it is not that people do not notice it, but they just do not want to think about it, hoping that their games of peaceful life with each other, their false assurances of friendship and love will protect them from an attack. No, they will not protect. And, it should be noted that the consequences of participation in such combat operations are the same for a major politician as they are for an ordinary, simple citizen—pain, psychological, physical, financial and other damages, etc., up to physical destruction. War is war.

Hence, the joke: "How do you feel, Joe?" – "Thank you, not well at all!" Life at war is difficult and it is even more difficult to survive. However, hiding in dreams is not just pointless, but extremely dangerous.

The whole question is not in ambition since we faced the fact that ambitions of the so-called "ordinary people" sometimes can be truly exorbitant, but in that what a person is capable of (himself, him personally). What information he has, what he knows how to do, how he does it and does he do anything at all. Since sometimes people behave quite strangely, for example: complain, get offended, upset, when someone reaches any success or surpasses them in their personal qualities in reality. These people are so used to being great and mighty in their dreams that any encounter with real facts that confirm that they are really not, make them very depressed and they feel hate towards other people. From our perspective, sometimes this reminds of a situation with a home dog, which lazily wiggles its tail all day in anticipation of getting fed and petted, or sleeps, pretending that it protects the house during the breaks, and with the return of the master after a hard day of work is puzzled and upset after getting only bones from a piece of meat.

Yes, usually no one physically destroys so-called "ordinary people" for their mistakes in communication, for misunderstanding and miscalculation of a situation, as it happens with, for example, politicians, officials, business leaders, etc. However, no one is immune to psychological destruction by another person or a group of people, which, as it is known, is followed by either physical destruction or, more commonly, physical self-destruction because psyche is primary. Usually this does not get taken into account at the beginning, and when it occurs it is usually too late to do anything about it—you, so to speak, got eaten. This society is a jungle, Ngorongoro, where predators hide behind the masks of innocent white sheep.

People, regardless of what social class they belong to, whether they like it or not are connected with other people. A lion's share of processes in which a person participates, is anyway contact with another person. Consider business, for example. At one time we encountered a strange, from our point of view, reaction of some not so major businessmen who were not immediately able to understand how the Catalog of human population is connected with sales of goods and services. While every business is, first of all, people. Development, production and delivery of goods are people, salespeople, managers, financiers and accountants, marketers—to people; and so on. Not to mention the service sector, which works strictly within a "person to person" system. Take humans out of a business system, and there will be no business in any sphere. Therefore, we do not quite understand why, for example, spend enormous amount of money in order to organize another talentless advertising campaign, or use incredible efforts to obtain acceptable terms in negotiations with a business partner, or fight for an "interesting client." Is it not easier to get information from the Catalog of human population about the person or people of interest, at whom these efforts are directed?

With the discovery of the Catalog of human population another myth is dispelled—the myth that a human and his "Soul" are unknowable. Today, one can just look in the Catalog of *Homo sapiens* and get information about a person that interests him or about himself because, as already mentioned, people are not familiar with their own selves either.

This is why any catalog is valuable. There is no need to engage in long and difficult study of a phenomenon or an object—just open a book and read about the features inherent in some form or phenomenon. This approach is even more useful with such a complex system as a human being.

Unknowability might, of course, be very romantic and mysterious, but it is extremely inconvenient in practical application. Hence, the question: does it make sense for a human to remain unknowable to his own self? Or, perhaps, it would be wiser to learn about your own self, and preferably before other people learn about you (and, not only relatives, acquaintances, and friends, but also competitors and enemies)? After all, representatives of the biological type *Homo sapiens* are still very, very far from Christian Love for each other.

PART 7

"IN THE BEGINNING THERE WAS AN IMAGE." WHAT ARE NATURAL IMAGES?

In the course of our research it became clear that all information about people in the Shan Hai Jing, which turned out to be the "Catalog of Human Souls," is encoded by images. An Image is behind every word of this "Divine Book" (except, of course, conjunctions, prepositions and parts of speech, which do not carry contextual meanings).

Andrey Davydov, the author of the discovery of the "Catalog of Human Souls," was able to find out that the record of all parameters of psychophysiology in human psyche consists of images. He discovered that an image is the basis of work of human psychophysiology on all six factors. Therefore, a complex system called "a human" is based and operates specifically on the basis of images.

An image, natural or artificial, works at the cellular level. This means that cell of a human body executes orders of images without question or discussion. Figuratively speaking, a cell is a soldier and acts according to the will of the Image-general without any reflections, discussions, and even more so bickering.

*****

Moreover, Andrey Davydov found out that an image as such is the underlying magnitude in this world.

Other ancient sources directly or indirectly confirm this. Not without reason the founder of Taoism, Lao Tzu, says in IV Zhang of Tao Te Ching (4-5 century BC): "An Image is the basis of all things." And, most likely, the early Bible instead of "In the beginning was the Word" stated: "In the beginning was the Image." After all, the Bible was translated from the Greek language, and in Greek philosophy 'eidos' (an image, a prototype, an archetype) does not occupy the last place. However, we do not undertake to state anything in this sense, as we are not experts in the study of the Bible.

However, it is not without reason that we can consider ourselves experts in that what is called by the word 'Image'. Also, we know that the results of our research, which clearly show that an Image is the basis of functioning of human psychophysiology, are confirmed by research done by geneticists. For example, some of them believe that communication between cells occurs on the basis of exchange of video information. And, video information is images because an image is really a picture.

*****

There are many images present in this civilization. They are on every corner, simply everywhere. However, all images that are created by people in this civilization are artificial images, whereas information about the human "Soul" is recorded exclusively by natural images. Natural and artificial images differ very seriously in terms of their impact on human psyche and physiology. The only difference between natural and artificial images is that natural images transmitted to a cell of a human body give unlimited health, youth, life, while artificial— slowly kill.

Consider natural images of human programs ("Soul"): each subtype program has its own unique and very rich set of images—a huge number of images in a variety of combinations. Combinations of images that human "operating system" is recorded with are unique. None of them repeat. In other words, each subtype has its own unique set of images that the program of this subtype is recorded with.

Our Creator made every human subtype a carrier of unique individual qualities ("Soul"), unique functions, unique way of evolution (Spiritual Development). And, all this uniqueness is encoded in natural images. A conglomerate of Images is the program of human subtype. A conglomerate of natural images is the "Soul." It is both an individual's personal "Soul," and the "Soul" of the whole subtype. The scientific name of this conglomerate of images is "Individual archetypal pattern." For those wishing to familiarize themselves with this in more detail, we recommend one of our scientific articles called From Carl Gustav Jung's Archetypes To Individual Archetypical Patterns.

Subtypes of type *Homo s.* have individual archetypical patterns of various sizes. Some human programs have around thirty images, while others have more than one hundred images.

However, it is worth noting that the quantity of images, by which human programs are recorded, do not make one subtype superior to another subtype or subtypes. Our research, in the form of decryptions of information from the "Catalog of Human Souls" and their comparative analysis, showed that as far as potential that "God" implanted in this or that subtype—all subtypes are equal. It is impossible to talk about competition between subtypes. It is the same as in the world of animals: a dog cannot compete with a bird or a fish, and lions are not superior to elephants. They cannot be compared because they are different. In the same manner, every human subtype has their own scope of functioning, their own range of abilities. Therefore, many of the existing ideas in civilization about superiority of one group of people over another is archaism, from the point of view that humanity is a type, in which there are 293 subtypes with different psychophysiological sets of qualities. Superiority of one over another (or groups) can only take place within a single human subtype, and not among people who belong to different subtypes. According to information from Shan Hai Jing, people can surpass each other only by one criterion: psychophysiological qualities, psychophysiological abilities that are achieved on the basis of knowledge from the Catalog of human population, but are not in any way connected with skin color, nationality, or association with some social group.

From our point of view, superiority among subtypes within the human society can exist only because some use knowledge about themselves and other people from the Catalog of human population, and some do not. However, we hope that this is a temporary phenomenon. This is happening only because today not everyone is familiar with the Catalog. According to our research, this is not the norm; competition among subtypes should not exist in the human society.

*****

As it turned out since an individual program of a human, his "Soul," his individual "Path" is recorded by images—recognition of these images from the "Catalog of Human Souls" and their study is the main task, goal of every human being. Because in the process of learning images of his subtype program a person learns qualities, functions that were intended by the Creator for him personally, and which were given to him through these Images.

Apparently, this was done so that every person could learn and study his own self through natural images of his individual program, implanted in his subconscious. "God" created Nature. On earth, it is mountains, valleys, rivers, seas, lakes, trees, grass, a variety of plants and animals, rocks, minerals, metals, birds, fish, reptiles, insects, flowers, sand, soils, and much, much more. Above it is clouds, fog, rain, wind, celestial bodies, stars, planets of the Solar System, and so on. All these are natural images. All that is left to a human is to learn by which images his personal program is recorded and to begin to study them in the environment. Due to existence of these images in the environment, a person can actually see, hear, feel, touch, smell, and touch them. And, through this, he can understand how these images live, function, look, move, smell, sound, and so on. Thus, a person receives knowledge on how he himself must live and function, what he should be to follow the "Plan of the Creator," his real self, his personal "Path."

Not without reason the Bible says that "God" created nature for a Human. Information from Shan Hai Jing confirms this. Not only this, but also that nature created by "God" is an ideal environment for survival of human species as far as composition of air, water, etc., food base, pharmacy, a vast reservoir of a variety of natural materials from which a person can build a house, make clothes, shoes, dishes, tools, and everything else that he needed in his life; nature also provides real living or non-living standards of how a person should be, how he must live. The Creator took care of his sons and daughters by creating all the conditions for them to live well and happily. All things were created by "God," and created in order for His children to conform to "His Divine Plan" and live the way He planned.

*****

If we use the language of science, natural images are etalon-carriers, examples of specific qualities present in this or that physical form. Their functionality can be translated onto functioning of human psychophysiology. Perhaps, this will be clearer through an example. Therefore, for example, to determine what psychophysiological qualities are inherent in a woman if one of the images of her individual program is a horse (Lat. 'Equus', the only modern genus of horse family Equidae, Perissodactyla order), it is necessary to learn the natural properties of this image. For example, we learn that a horse is a perissodactyl animal; weighs up to 300 kilograms; polygamous; herbivorous; lives on the plains; gallops with a maximum speed of 60 kilometers per hour. (This is far from being complete, but this is only an example.) This information is then transferred onto a person on all six factors. As a result, we get the following information about this woman: that she thinks clearly and expresses her thoughts directly; since horses have a strict hierarchy of females, she is a pronounced leader; by nature she has an athletic body, and, as a rule, it is well trained (or looks like it is, even if she does not train on purpose), has a moderate fat layer, a tendency to run (although she is not a sprinter, but a stayer); in terms of nutrition, she is a vegetarian; in the sexual factor, she tends to have a variety of sexual partners, and at the same time she is a very jealous person; emotionally testy, expresses her emotions openly and unashamedly; as for the environmental factor, she is not a good housewife—her home, will be about as clean as a stall of a horse...

This is about one hundred-thousandth part of what just one image of a person's program can uncover about that person. However, a human is not an animal that has only one image at the base of its functioning. As already mentioned, the number of images in some human programs might exceed one hundred, and each natural image is first of all looked at in terms of its natural analog.

The Creator left this method of learning his own self to a human, so that a person could clearly see the difference between his understanding of natural images (imagination) and real natural objects. After all, *Homo s.* has an ever-present desire to imagine, to understand objective phenomena as he wants, in a way that is comfortable, pleasant to him. And, that is fraught. As a result of such "understanding" he, figuratively speaking, will try to treat a bear as a teddy bear. However, our Creator has foreseen this. Possibly, he created two of every creature, so that people see the objective natural properties of this or that natural image that is within himself, and study.

PART 8

"THE WAY TO SPIRIT" IS TO STUDY AND TO LEARN

Natural images are etalons—etalons of personal qualities of a human, etalons of how he must live. As mentioned above, "God" created these etalons as objects of nature, which humans must emulate and conform to, and left the source (the "Catalog of Human Souls") from which one can learn about his personal etalons. In light of Andrey Davydov's discovery of the ancient source of knowledge of human "Soul," it became clear why all ancient cultures had a ritual to worship natural objects such as animals, plants, etc. It turns out, our ancestors studied the "Divine Instruction" on what they must be and how to live in order to be creatures as powerful and with incredible opportunities as He, the Creator. Since, according to "His Will," fixed in the "Catalog of Human Souls," only natural Images, those images, which have been created by Him, and not the ones invented by people (and widely presented today in film, literature, the Internet, etc.) are the etalons of how people must live. However, at some point humanity violated this "Divine Law."

As documented in history, with disappearance of the Sacred tripods, humanity lost the source with descriptions of individual "Paths to Spirit", "Paths to God," the Instruction to themselves left by "God." There was nowhere to get information on what they must be, how and what for they must live, and so people (or the one who hid the Catalog, and imagining himself already equal to the Creator) started to make-up artificial images; since a human being has been created in such a way that he cannot exist without an "operating system." Thus, humanity came into total contradiction with the "Divine Will." This period in the history of human civilization is called by different names, but the essence of processes that occurred is well demonstrated by titles like the "Golden Age" and the "Iron Age." Since with the loss of the "Catalog of Human Souls" the quality of human life has deteriorated significantly.

And, it is very likely that if people had qualities and abilities truly equal to the Creator's, all would be well with creation of artificial images. However, humanity has begun to do this before reaching the status of "Image and Likeness" of the Creator. Due to this it is unable to create images that are equal to natural images that He created. As it is known, an analogous situation exists in relation to other artifacts. To date, any human-made product is not equal to the level of creation, the quality of creation by nature ("God"). Quality of nature's creation is always superior to quality of human's creation. For example, synthetic leather by its properties is significantly inferior to natural; synthetic diamonds are not comparable to natural diamonds; this list goes on and on. Today, chemists are not able to artificially create even a relatively simple essential oil of lavender (only 250 components) that would be equal to natural.

*****

At the same time, according to information received in the course of our research, the "God"-given potential implanted in a Human is huge, and that is really impressive. Potentially, a human is able to have colossal power, abilities unprecedented from the point of view of this civilization; in short, all that what is common to put into the category of "supernatural" in this society. In fact, according to information from Shan Hai Jing, these are not superpowers, but are ordinary, normal human abilities. Which however, may be "in a sleep mode" throughout the whole life of a person and never manifest. Due to the fact that these abilities do not open up without special efforts, it is possible to activate these treasures in one's own self only if one knows his program from the "Catalog of Human Souls," realizes it, studies, learns.

Prior to realizing the full spectrum of abilities implanted by nature, any living natural object must go through the path of growth, development, learning—a seed does not have the same properties as a tree, a chick does not have possibilities of an adult bird. A kindergartener is simply unable to write a doctoral thesis; instead of complex mathematical calculations that a professor of Mathematical Sciences can do, a first grader barely adds two plus two. However, this is not because they are idiots, they just need to grow up, go through the path of learning in a school, then at a university, and so on. Instead of a visual art piece that a professional artist could create, a child draws some doodle, but not because he has no talent in art, but because he needs to study.

As far as more complex things than math or fine art, such as acquisition of qualities of the Creator—development and growth in the case of a human being occur only on one condition: if a person knows and implements that program, which "God" implanted in him. And, studies his entire life, but studies not just anything.

Perhaps, an explanation is required for a person who lives not on the basis of his natural Images, but on the basis of artificial ones. The word "study" does not mean getting an education at an institution; it means self-knowledge and mastery of one's natural qualities, as realization of his natural talents. The word "work" does not mean to dully go to work every day, even if one diligently performs his duties, and it does not mean to rush somewhere with a goal of "making a few dollars." According to the "Divine" source, "to work" means to work on one's own self, his own personal qualities, to work in the process of self-improvement. Only then, as it is written in the ancient book called Shan Hai Jing, through his qualities, a person can really become equal to his Creator. Other paths simply do not exist in Nature. The clearest examples are results obtained by humanity living based on lack of knowledge of themselves.

As Shan Hai Jing clearly states, knowledge and development of a Program is the most important thing that a person has to learn how to do and study. Therefore, it is safe to say that with the existence of an individual "Path to Spirit" for each person, there is one "Spiritual Path" for all humankind—to study, to learn.

And, here is the evidence. The first part of the "Catalog of Human Souls" contains models of psychological structures of *Homo sapiens* who have not yet fully moved away from the animal world in their development. Therefore, among the images there prevail images of animals, plants and the like—stones, metals, rivers, lakes, seas, plains, mountains, and forests. However, the next parts of Shan Hai Jing describe other images, which indicate transition of a human to new levels of development. And, spirits, deities are always present among images of these levels of programs. Any person can verify this. One only needs to open Shan Hai Jing in any language, not necessarily an original, and compare descriptions in the Book of Mountains of this source to the remaining chapters that follow.

However, in order for a change from one natural program to another (a program of another level) to occur, a person must first fully, "from A to Z" realize the initial program with which he was born. And, he must realize it not in his fantasies, but in real life—make himself exactly the way he was created. "God" left the "Catalog of Human Souls" to people with parting words—study hard and work hard. Studying hard and working hard is the essence, a summary of characteristics of the image called "Spirit," which according to the text of this Book, appears in a human who has realized his primary (first level) program. As we have found out, this is the methodology of what is called "Acquisition of Spirit," Spirituality.

According to descriptions found in Shan Hai Jing, the next level of "operating systems" described in this ancient treatise is a more advanced stage. It is accompanied by liberation of a human from former psychological structures and achievement of consciousness of a higher level of development. From our point of view, this is the real transformation of an individual, true human evolution; rather than that what is still sometimes seriously called by the word "evolution"—evolution from an ape that *Homo sapiens* never was.

The "Catalog of Human Souls" describes several levels of development, unique for each subtype. However, it does not make sense to discuss them in this very brief overview. Especially since any development always assumes full completion of the initial stage first, and jumping over steps like on regular stairs simply will not work. We suppose that right now it is enough to know that descriptions of individual "Paths to God," individual "Paths of Spiritual Development" have been found, and that any interested person can start using this information right away.

And, for those wondering what "superpowers" this method of self-improvement and Spiritual Development of a human uncovers, we recommend reading about the abilities and capabilities possessed by gods and spirits in various mythologies. Since, from our point of view, ancient myths are not fairytales, but fixations of real events that occurred. Our ancient ancestors lived in very harsh conditions and had to work hard day and night to earn their living. When, and most importantly why did they engage in such an unserious thing as creation of images of fantastic animals, spirits, deities? As much as the descendants of the ancient people want to feel superior in their level of development, we are of the opinion (and we have good reasons) that the ancient were not so naïve, unpragmatic, and careless to create, store and transmit empty knowledge to their descendants.

*****

For each person a clear knowledge of the "Will of the Creator" with respect to his own self provides huge benefits, opens incredible opportunities. Even if we talk about the most plain ones that relate to daily routine, even then instead of continuing to exist at the level of a half-animal, trying to fit into conditions and roles proposed by secular society, breaking and deforming oneself—now any person can be his real self, and live his own life. And, a life that is not invented, directed by someone. After all, people do not even know that they are not the ones who invented artificial images, which they think is their precious individuality. These images are actually invented by secular ideologists to make people manageable and comfortable for use.

Such artificial formations as "duty", "guilt", "shame", etc., which were invented by ideologists of secular society in order to have control over an individual, are also losing power. Manipulative threads, by which, figuratively speaking, a human is pulled like a puppet—will be permanently torn. However, that will hinder survival of an individual in the environment by a lot. The Creator created all natural programs in such a way that any person could fit into the world well, both in nature and in society, and without loss of strength, as it happens now, be really powerful, healthy, talented, brilliant, harmonious and happy. In this case, an individual will consider society as a free supermarket. However, that also must be studied and learned. The main thing is to have the keys. (We will discuss this point in greater detail in another book—Society As A Free Supermarket.)

However, most importantly, a person needs to learn in order to one day rise to the level of his Creator. This purpose is implanted in every human being as the goal of his life, for any individual, regardless of subtype. This is the purpose of existence of the biological type *Homo sapiens* on earth, and this is recorded in the "Catalog of Human Souls."

*****

Discovery of the Catalog turns an ordinary view not only on the entire ancient mythology (and many other things) upside down; for example—the ancient pyramids. As already mentioned, the ancient source (the "Catalog of Human Souls") of knowledge about a human and his Spiritual development describes 293 subtypes that exist on the first level. However, after the descriptions of intermediate levels, at the end of this treatise there is a description of only one program—Human. No subtypes. Other material sources in this civilization show the same. As we were able to find out, this exact principle has been fixed in such a structure as a pyramid (while scientists are still trying to find the meaning behind pyramids).

Pyramids as architectural structures exist not only in Egypt or Mexico. Pyramids as built structures can be found in different cultures: Mesopotamia, Sudan, Nigeria, Greece, Spain, China, Korea, Central America, North America, India, Indonesia, Peru, and so on. Pyramidal buildings also existed in the Roman Empire and Medieval Europe. From our point of view, this once again confirms that at some point in human history the Catalog of the human population was present all over the world. From our point of view, researchers of pyramids are absolutely correct in their observations—this very common form of architectural structure rarely carries a utilitarian function.

Our research suggests that pyramids are not just places of burial or temples. We discovered that a pyramid is a structure that fixates in stone for centuries, for many generations the main treasure and "compass" for human civilization—the Catalog of human population. Any person can confirm this right now. To do this, simply open Shan Hai Jing (Catalog of Mountains and Seas) in any language and look at the structure of this source. (Although one certainly should have the full version of this book, as it is sometimes sold in sections, for example, only Catalog Of Mountains.) By going through this book from beginning to end, it is not difficult to notice complete similarity between the structure of descriptions of human programs in the "Catalog of Human Souls" and structure of a pyramid: a construct as a base with four directions (in the Catalog these are chapters titled Eastern Mountains, Southern Mountains, Western Mountains, and

Northern Mountains) and the center (in the Catalog titled Central Mountains), and the book ends with only one description—the description of a being called "Human," Human as likeness of "God."

Therefore, ancient pyramids is a fixation (for centuries) of the principle of development of the human biological type, the principle of Spiritual Development: if an individual studies and learns, he leaves the zone of varieties of human subtypes that are at the level of development of half-animals, and becomes a creature called a "Human" ("God," Creator). Anyone who develops in this way, using the language of the Bible, makes himself the "Image and Likeness" of "God."

*****

The principle of the method of development for humanity as a single species (using the terms of religion—the "Spiritual Way", "Path to Unification with God") is fixed in another ancient structure of this civilization—in the ancient Egyptian Sphinx. Perhaps it is not by accident that mythology of different nations is so full of chimeras that combine features of several different animals in a single image.

By its form the Sphinx as if tells people that unlike animals, a human has a more complex structure that is recorded by more complex subtype programs. And, since these programs carry by far more qualities, functions, and abilities than programs of animals, they are recorded not by one image, as in the case of animals, but by many images. From out point of view, this principle is clearly shown in a Sphinx because it is a chimera, a collective image that has properties of several animals: a human head, a lion's body, a tail of a snake. From our point of view, the solution to the riddle of the Sphinx is that in the case of *Homo sapiens* under a human form hide a variety of images, including images of animals. As studies of Shan Hai Jing clearly show, a chimera is a principle of human structure. From our point of view, someone has built the Sphinx a long time ago specifically to demonstrate this principle. This is why the Sphinx has a human face—to let people know that this principle applies to them, to *Homo sapiens*.

*****

In conclusion in regard to the main objective of growth and Spiritual Development of a human, it is necessary to note once again that those *Homo sapiens* representatives, who have no goals to make themselves in the "image and likeness" of "God," who are not familiar with their "Soul" and with the "Catalog of Human Souls," who do not wish to study and work (primarily work on becoming consistent with the will of the Creator) will never have any qualities of the Creator. Do not even dream of it. And, in fact, this has long been confirmed by material results obtained by the people of this civilization.

PART 9

CÆSAR

WHAT ARE ARTIFICIAL IMAGES?

As already mentioned above, "God" created a human in His Image and Likeness, and implanted potential of a Creator into each person. And, gave everyone the freedom to choose exactly how and in what direction to use their existing abilities. The "Divine Will" in respect to a Human was recorded in the "Catalog of Human Souls." However, it so happened that humanity lost this Catalog, and began to use the "God"-given potential in a completely different way. In the direction opposite of the "Divine Plan," people began to create artificial images.

As a result, throughout all these years civilization produced an incredibly big mass of artificial images. And, every epoch, each culture created and creates its set of images: there were Apollo or Odin, then Zorro, Batman, Shrek, etc. This civilization has come up with a lot of modifications of artificial "software" without coming close to even approaching the question of every person having natural "software" ("operating system") given by nature at the moment at birth. On the one hand, this is explainable since people had to live on some basis because they differ from animals in that their "operating system" of the "Soul" does not work "automatically," as in the case of animals, and must be studied, comprehended and mastered. But, on the other hand, artificial "operating system" differs from natural like papier-mâché or plastic apples differ from natural apples.

Civilization involved in creation of artificial images, figuratively speaking, forces people to eat plastic apples instead of real ones. It is not difficult to guess that such "diet" leads each person to imminent death. No one has ever been able to remain alive and healthy while drinking gasoline instead of water, and replacing all natural foods with synthetic.

This, so to speak, "plastic diet" leads to an enormously rapid wear of the entire human psychophysiology, premature aging and, as a result, death. However, the result is extended through time, and it is similar to the effect of a person taking daily micro dozes of poison.

The potential implanted in a human is realized by 5% at best due to these "apples" (artificial images).

*****

The essence of the phenomenon called "artificial images" can be expressed differently. For example, in the language of religion, it is those evil demons and devils that possess, and force people to act contrary to the "Divine Plan." As a result, they eventually force a person to engage in self-destruction. In other words—they actually kill. If we explain the same phenomenon using the computer language, this is the worst of viruses that can infect a computer. For a while, this virus just disrupts normal algorithms of functioning of a system, but, as a result, sooner or later these violations make a computer completely unusable.

Translated onto human psychophysiology, artificial images as a "virus" cause the human organism to work incorrectly, not at full capacity, cause mistakes (including fatal ones). The practical results of these failures are a variety of misfortunes that happen to a human: failures, all sorts of problems, illnesses, and old age. Sooner or later, due to these violations, each of which break the system called human psychophysiology more and more, a human goes out of order, in other words—dies.

In the case of computers, probably in some cases a good specialist can fix a serious failure. However, in the case of a human, this occurs without a possibility of recovery. Thus, it was stated above that artificial images kill; slowly, but daily and inevitably.

Any artificial image (an image that was not created by "God," but invented by a human) disrupts human functioning as a living system so powerfully that sooner or later this system gives under. And, in this civilization it is observable everywhere and can easily be noticed. People rarely live more than 90-100 years. And, this is at best since human life is usually much shorter. Meanwhile, physiologists state that by nature human body is designed for a youthful and healthy life of a few hundred years. And, the Bible says that people who follow the "Divine Will," who realize the "Divine Plan," like Noah, lived up to 800, 900, 1,000 or more years. From this it is easy to conclude that all human troubles, sickness, misery, and as a consequence, reduction of length of their lives is the work of those "evil demons," which are actually produced by people themselves. This is "God's Punishment" of those who digress from His "Divine Will."

The Creator created a human as such a powerful being (potentially) that human psychophysiology can work for some time even when it is infected by a "virus." No animal would survive in such living conditions that people make for their own selves and remain alive in them for a while. "God the Father" took care of a human by implanting natural images as a basis for his functioning, and thereby provided a huge strength margin to a human being as a living system. However, one should understand that if he is still alive, it does not mean that he lives right. Moreover, it does not mean that he or she will live forever.

People should probably also know that it is not the "all-seeing God," who does the process of tracking of which images people base their lives on, but it is one's own organism, on the inside—that is how it is programmed.

Nothing can hide from this mechanism; it cannot be fooled or cheated. And, people have to pay the highest price for their mistakes in regard to their psyche and body. Sometimes the price is very high. This refers to biological death of an organism.

*****

It should be noted that according to our research of Shan Hai Jing and other ancient books related to it (we do not want to mention those books at this point), death is not a mandatory event for a human. Even more so, death is not natural at an age that in civilization is considered to be death as a natural order of things.

According to these sources, death is the result of a human mistake or mistakes, which caused a fatal damage of him as a living system. And, if a person stops making such frequent and fatal mistakes, he can live as long as he wants. (Shan Hai Jing mentions such long lifespans that it is not worth to tell modern people about them because they will sound too fantastic to them.)

As stated above, a "virus" called artificial images makes people make fatal mistakes. Errors that led to death of an organism are actions, behavior, human decisions. It is such a wide range of causes and effects that a detailed analysis in the format of this text is not possible. However, in a nutshell, people die not according to the laws of nature, and certainly not on the principle "God gives, and God takes away," but above all, due to lack of information about themselves, about others, about the world, about nature.

Also, from our point of view, religion is absolutely correct that this is occurring due to deep ignorance, laziness, stupidity, insistence on continuing to be stupid, unwillingness to work and learn; in other words, due to unwillingness to lead a life of a human instead of that of an animal. However, absolutely each one of qualities mentioned above are qualities that a person receives by creating or using artificial images. There is no other reason. "Devils", "demons" are not something that exists externally and comes to a human from the outside. "Devils", "demons", "evil spirits" are that what each human creates himself, and that what lives inside him.

As artificial images, these "evil spirits" appear and multiply due to human pride, laziness, and ignorance; because a human imagines himself equal to "God," and on that basis believes that it is not necessarily to study, to learn, to improve. He thinks that he is already great, perfect and mighty. Blinded by this pleasant illusion, a human seems to become blind, and does not notice how his own demons-monsters gnaw him, that they work toward his destruction.

If we draw a parallel with the idea of a sin that is present in religions, then without doubt we can conclude that creation of artificial images by people instead of use of their natural "operating system" that "God" gave them, is the "Mortal Sin" from which all human troubles and misfortunes come from, death included.

*****

Perhaps the only cause of death that cannot be attributed to human error is violent death (murder). In all other cases, people themselves are causes of their death, as well as diseases and other misfortunes that happen to them. People who live based on artificial images become dull and useless: they do not want to know and understand anything; do not want to do things right, but only as they want; do not want to take care of themselves, their lives; do not want to study themselves and the world; they are not interested in anything, do not want anything; they do not strive for anything, except for pleasure and lounging; they are too lazy to take care of themselves, their own health; they violate natural and moral laws; they engage in self-destruction as well as destruction of everything around them, and much more.

However, the worst thing for a human is lack of information. Since when a person does not have required information about the external and internal world, he begins to fantasize, imagine. Artificial images are a fantasy. Including human fantasies about their own selves. And, this is the worst thing that can happen to a human since that is the reason why all sorts of misfortunes happen to him. Often people are so involved in their fantasies that they simply do not want to see the surrounding reality, even when information is literally right under their noses. A human thinks: "So what if I am wrong, but I will continue to think and do what I want, as it is comfortable and pleasant to me." Each person certainly has every right to hold this position. However, from our point of view, this is typical suicide, and in a perverted form; a personally programmed euthanasia of oneself: "Let me die from this, but until the moment of death I will feel good, pleasant." "God" gave life to a human, and not for one's pleasure, enjoyment and convenience.

Also, people should not forget that Images are not toys. People invent monsters and become these monsters themselves, although they do not monitor and regulate this process due to their ignorance. However, cells of a human body do not account for ignorance. Cells do not argue what will be better for a person, as that is not their function. They simply embody (at the physical level) that image, which a person has chosen for himself.

This is how an image works; regardless of whether a person knows about this or not. As a result, a person actually becomes those images, which he has chosen. This is the free will given to him by "God." However, it implies responsibility as well, so each person always pays for his choice.

As it turned out, if any artificial image is a demon, an evil spirit, a devil, then it is not difficult to imagine whom modern humans turn themselves into, and where violence, destruction, death, drugs, war, and disasters actually come from. People attribute occurrence of these phenomena to anything and anyone, only not to themselves, their actions, and their decisions. However, people should know that in fact they are the ones who cause all this abomination.

Moreover, absolutely every person adds to this, so to speak, "bank of abominations." No matter how holy, sinless, innocent, and pure he considers himself. This is the case only because he lives not on the basis of an "operating system" created by "God," but on the basis of artificial images, which turn him into a monster in a human form. All people in this civilization are like that because an artificial "operating system" makes them so, and in a civilization where there is no "Catalog of Human Souls" there is no alternative.

*****

Unfortunately, all of the above is not just another horrifying fairytale of non-fiction, as readers would like to think. All of this has scientifically proven evidence: all mutations in nature are subject to destruction. Artificial images instead of natural ones that are used as etalons of individuality by a human, lead to the state of a mutant. Therefore, nature shortens lives of people who live on the basis of artificial "operating system" to a minimum, and then at some point turns on the program of utilization, disposal. Nature does not need people who transform themselves into mutants through their mad fantasies, who do not fit into the normal environment in a natural way. Moreover, they present danger to the rest of "God's Creation." From our point of view, the natural mechanism that destroys mutations is described by the saying: "God gave, and God took." Only instead of the word "God" it would be more correct to say "nature." But, on the other hand, it does not matter since "God" created nature.

PART 10

"DEAD SOULS": HUMAN MUTANTS

> "You have a name that you are alive, but you are dead."
> Apocalypse (3:1)

Despite the fact that in the modern civilization we could not find a single detailed scientific description of the phenomenon called "guism" (we are referring to our research as a decryption of another ancient manuscript, in addition to Shan Hai Jing, details of which are not possible to cover in this overview), there are enough references to this in Christian religious sources. Meaning that living people with a dead "soul" exist in reality: "As many of the living are dead, burying as into a grave, his soul into the body...", "Many in the living body have soul of the dead, buried as in a grave.", "Soul departed from its virgin nature, dies..." (Rev. Abba Isaiah (82, 230-231). We fully agree with this because, according to our research of the "Catalog of Human Souls," as soon as one deviates from realization of his natural program, he becomes a living human, who's "Soul" is dead.

However, we would like to remind of the fact that we are not theologians or religious scholars; therefore we cannot claim to know the content of all religious sources available in this civilization. We are referring to those, which we found in Christianity. However, if we consider these sources, then the essence of references to living human beings with dead "souls" is convergent with our conclusions based on research of Shan Hai Jing, although with minor corrections. The Christian sources say that the "Soul dies when God leaves it," but from our point of view, it would be more correct to say that the "Soul" dies when a human leaves "God." Or, to be even more precise, when a person leaves his "Soul" as his own self, his real self, his natural self; leaves himself as a representative of his subtype with "Soul" that the Creator created for him.

Perhaps, representatives of the modern religious confessions, institutions, organizations, etc. might also be interested in our research of the phenomenon of a dead "soul" in a living body. We will try to briefly explain the essence of this phenomenon.

It should be noted immediately that the following is neither our hypothesis, nor our conclusion, and certainly not our subjective opinion in regard to a human whose "soul" is dead. The following characteristics of a person with a dead "soul" were taken from another ancient book that we have researched. Again, in this case, we are only translators from one language to another, and we adapted this information for understanding of a modern human, without any of our own additions.

Without going into details on how and where the following information was obtained, we only present excerpts from descriptions of this phenomenon.

A Mutant: A Living Human With A Dead "Soul"

We suppose it is appropriate to begin the description of this phenomenon with a definition that we made based on analysis of an ancient text that we discovered: "A human with a dead "soul" is someone who has a narcissistic, self-fertile intellect, aimed at achievement of absolute subordination of all systems of his/her own psychophysiology to himself/herself and relegation of these systems to a state of secondary segments."

From our perspective as researchers, this phenomenon does not belong to the category of some special phenomena, nonsense, or rare facts because it is widespread, available for observation (self-observation) to anyone.

Also, it is absolutely clear that this phenomenon does not belong to the category of natural phenomena. From our point of view, it is 100% artificially created, civilizational.

A human with a dead "soul," or using the scientific term—a mutant, is a product of modern civilization. This product results from a creature that came into being as a human. Since this civilization has no basis in the form of knowledge about the nature of human psyche, it breeds, using the computer language, artificial "operating systems" for people, which get recorded by artificially created images and perceived by *Homo sapiens* as their own, personal "operating system," as the true "Soul." However, since the Creator created a human with a completely different "operating system," recorded not by artificial, but by natural images—from nature's standpoint he is nothing but a mutated individual.

This type of mutation (from Lat. 'mutatio' – alteration, change), to put it in scientific terms, has psychical pathogenesis. Since psyche is primary, the pathological changes in this case spread over the entire psychophysiology of a human (on all six factors), unlike most biological mutations, which do not touch upon phenotype as a set of characteristics inherent in an individual at a certain stage of development.

Similar to mutations that occur in a living cell, which are known from biology, the considered type of psychical mutations appears as a result of impact of artificially created images on an individual. The consequences of this are comparable to consequences of biological mutations, in cases when mutation affects functioning of a cell, which triggers nature to launch utilization mechanisms, and the cell gets destroyed. Only in the case of a "virus," which are artificial images—an individual, as a cell of a single organism called "biological type *Homo sapiens*" gets utilized.

In the case of the considered phenomenon, mutation indicates pathological changes in the functioning of a human program, which lead to breakage of all systems of human psychophysiology.

*****

We will start the description of this phenomenon by explaining the terms that were used in the definition. The word "someone" indicates that a person with a dead "soul" is neither an Individual, nor (and especially not) a Human.

By the word "narcissistic" we mean a feeling that is universal for all people with dead "souls"—a feeling of one's own greatness, uniqueness, value, significance, which in most cases is not confirmed any real qualities or actual state of things. A human with a dead "soul," whoever he is in reality, invariably suffers the illusion of self-importance, possession of qualities that are not actually present (for example perfection, omnipotence, immortality, and so on).

Narcissism, hypertrophied self-centeredness, feeling of being the "center of the earth" around which everything revolves is inherent in people with a dead "soul." A person with a dead "soul" is convinced of his exclusiveness. One of the characteristics is that he considers himself already a Human (a being equal to the Creator), the "Crown of Creation", "King of nature." Although by his true qualities, he is just an involuntary living machine that is not far from the level of an animal.

Next, we used the word "self-fertile," which meaning is as follows. In the works of modern researchers in the field of biology, the term "self-fertile" means the ability of a plant to produce normal, healthy offspring after self-pollination. In the case of a human with a dead "soul," perhaps it would be more correct to use the word "self-sterility" since the fruits of his activities are hardly possible to categorize as normal creative products. Products of autogamous, "self-pollinating" human intellect are extremely anti-natural and ugly.

On the other hand, the word "self-fertile" successfully reflects properties of people with dead "souls." Throughout their lives, figuratively speaking, they boil in a soup of their own impressions, fantasies, and illusions. They live not in reality, but in illusory worlds that they organize, and which have little relation to reality. The reality is far from them, due to their own decisions, and they struggle to keep reality out of their lives. They resist, so to speak, "natural pollination," meaning any information from the real world. Self-fertile intellect is wild imagination, fantasy; constant departure of an individual into the space of his personal mental constructs. However, regardless of the fact that in psychiatry this is considered a mental disorder—all people live this way. Perhaps, it is not without reason that the following joke exists among psychiatrists: "All people are crazy, only some simply do not go to doctors."

Every person with a dead "soul" carefully protects his "virtual world," believing it to be his real world, the world of his "soul," his own self. For a long time, and despite the fact that reality regularly invades his life and destroys this flimsy "house of cards" that can fall apart even at the slightest breeze, a person with a dead "soul" habituates himself to not see, hear, and understand all that what could destroy this phantom. And, in the end he or she accomplishes this, although this negatively affects both: cognitive abilities, and the quality of work of his or her intellect in general.

A person with a dead "soul" is therefore classified as one who has a "self-fertile", "self-pollinating" intellect because he actually shuts himself off from information from the outside, and especially carefully protects his "mind" from intellectual schemes that do not support his own, from any information contrary to his "vision of the world." Although he cannot have any "vision" as he is blind, deaf, and so on. He is unreceptive to anything outside; everything is perceived with great distortions, and this is pathological as far as the normal operation of perception as a function of psyche. The pathology is that any person with a dead "soul" strongly believes that the world is as he imagines it.

Another characteristic of a person with a dead "soul" is that he does not wish to change his opinion about anything whatsoever, even if there is fear of physical death. (Doctors often face this as well.) A person would rather sacrifice his own health, well-being, comfort, money, relationships—anything at all, just to not bring into his once established views of the world anything that could break them, and even more so destroy or change radically. From psychical standpoint this is a truly dead person because only the dead stiffen in one position and remain in it forever.

Also, our definition stated that intellect of a human with a dead "soul" seeks to bring down the rest of the systems (psychical and physical) to a state of secondary segments. We think it might be easier to explain this using an example. Assume that "intellect" of a person decided that his host or hostess must become a vegetarian. And, it produced a corresponding attitude: "Starting tomorrow, I will eat only vegetables." Although one of the images of the individual program of this person could be an exclusively carnivorous animal that eats only meat. In nature, such animals are known to exist.

It is not difficult to predict the results of this experiment. First, this person will have a sharp decline in physical and psychical activity; he will experience weakness, flaccidity, depressed emotional state and mood, etc. However, his intellect will not take this into account; at best, it will find an excuse not related to reality, and at worse will say to itself some nonsense like: "You just don't have the willpower! You must do this, get a grip, do not give up!" After the initial results of such a seemingly harmless experiment, the person's organs and systems will start to malfunction because they no longer receive the necessary substances from food. And then, he is on a road straight to physical and mental diseases. However, a human with a dead "soul," one who, figuratively speaking, carries a "demon's head" on his shoulders in the absence of normal intellectual activity, is unable to see and understand the basic interconnections of cause-and-effect. Therefore, he will be unable to conclude that, for example, an urgent recent surgery (for example, related to digestive propulsion), or the fact that that he was expelled from an institute due to poor performance, or that he was fired from his job because of his mistakes and negligence, or that his wife left him because he began to have problems with potency—are all in fact direct consequences of his decision concerning rejection of meat.

The example above was provided in regard to the nutritional (dietological) factor, but intellect works exactly in the same way in regard to all other factors: physical, emotional, sexual, environmental. Although normally all six factors of human functioning must work on a parity basis. By nature any one of the factors is not the main, the primary one, and needs of all factors must be met. However, the intellectual factor of a person with a dead "soul" runs the work of his entire organism. (Well, at least tries to.) On his own, autocratically his "intellect" decides how his physiology, nutrition, emotions, and so on must work; "intellect" decides which needs of other factors to meet, and which not to, which functions to realize, and which are not necessary to realize.

"Intellect" of a person with a dead "soul" claims control over all of his other natural factors: physical, nutritional, emotional, sexual, and environmental. By organs and systems of the human body know from anatomy, we are referring to the circulatory system, cardiovascular, respiratory, integumentary, digestive systems, bones, kidneys, liver, and so on. This "intellect" tries to control the work of psychophysiology, direct it in whatever way it excogitated, and that causes colossal harm.

However, a human is constituted in such a way that his intellect, in spite of all the attempts and highest ambitions, cannot replace a program. The most it can do is deform it, break the natural mechanism of a human. And, this is very easy to do: just stop satisfying natural needs (meaning needs of the natural program), and satisfy fantasies instead. It is not difficult to guess what this will lead to. If for a moment we assume that some animal, for example, a snake, tries to do something like that, and instead of meat begins to eat grass, nuts or seeds, it will die quickly. If any plant is put into unsuitable soil, in wrong temperature, light or other conditions, and, in addition, for example, hydrochloric acid is used to water it instead of water—the plant will not live very long. The same occurs with a person who satisfies his fantasies instead satisfying the natural needs of his psychophysiology, and this is the reason people live such short lives. Only, unlike it is with animals and plants, the result in the form of biological death of a human is stretched in time. On the other hand, in this civilization a person does not have any other way because it is impossible to know the true needs of his psychophysiology without knowledge of his program from the "Catalog of Human Souls." As to why, was already explained above.

"Intellect" of a human with a dead "soul" usurps authority over his entire psychophysiology, and in the end always successfully—the person dies. And, as it was stated in the definition, only because schemes, opinions, views, desires formed in the brain of an individual by culture, upbringing, and education (this process is described in more detail in our book titled Ahnenerbe—Your Killer Is Under Your Skin) purport to substitute the program. As a result, natural algorithms get relegated to secondary, and practically get destroyed together with their carrier.

Very briefly, below are characteristics of a person who, according to identified criteria, is a person with a dead "soul."

"Demon's Head"

Intellect of a person with a dead "soul" works, to put it mildly, in a very peculiar way, very different from the natural way in which it must work. Intellect infected by a "virus" on the basis of artificial images (using the terms of religion—"possessed by evil spirits") works in an unnatural manner. It is narcissistic and self-fertilizing, and so, by and large, it is incorrect to call it intellect in the full sense of the word (this is the reason we think it is more correct to put this word in quotation marks). "...Intellect is killed in relation to eternal life..." That is why, if we use the language of religion, it can be called "demon's head."

The main vocation of a "demon's head" is construction of yet another perversion. Especially, since in this civilization for some reason this is regarded as genial. (Although this has nothing to do with genius.) By "perversions" we mean perversions in terms of nature, its system, and the human system in particular.

The first and most important perversion is that a "demon's head" has an ability to produce schemes, which replace work of the biological program of bio-robots of type *Homo sapiens*; and that is—creation of artificial images.

This activity of "intellect" of an individual replaces knowledge about himself, about the structure of his psyche, its content, and algorithms of work of his own physiology. Since, as already mentioned, unlike animals, biological programs of which work completely automatically, normal functioning of the biological program of *Homo s.* requires participation of his intellect. Therefore, by nature intellect has access to certain parts of human psychophysiology. However, psychophysiology requires only some participation of the intellect, and in the case of a human whose "soul" is dead, the intellect tries to fully replace work of physiology with "virtual life."

(More details on this phenomenon are described in our article called Virtual House Where Hearts Get Broken.)

It should be noted that intellect is given to a human by nature in order for him to communicate with his "operating system," and this is required for normal functioning of psychophysiology, for processes of continuous growth, development, and gradual changes. In other words: to solve problems, which representatives of other biological types do not have, only humans. However, intellect of modern humans does not work. The reasons were already specified above. According to psychologists themselves, modern science does not know for sure what human "soul," psyche is, what its structure is. The place of psyche in a living system, its core functions are also unknown. The so-called psychophysiological problem (the essence of which consists of incomprehension of cause and effect relations between psychological and physiological processes in an organism in accordance with psychical and physical phenomena) is not resolved to this day. In such a cultural environment, characterized by lack of answers to basic questions that people have, humanity is trying to survive for many centuries.

Normal intellect turns into "intellect" because it has nowhere to get information on how it is arranged as a living system, how it should function, for what purpose it was created by the Creator. Also, it is unknown how one's intellect should normally work, what functions it must perform, which areas to stay out of, and so on. As a result, a modern human has no intellectual activity as such; there is only imagination. Although normally thought processes, not fantasies should prevail. And, thinking occurs only based on one scheme: information gathering, analysis and conclusions. Since the required information is not present, there is no thought process. Based on this it is possible to conclude that intellect of modern *Homo sapiens* practically does not work, it is turned off. This especially manifests in cases when it comes to the person himself, work of his psyche and body, as the first mandatory item of thought processing is impracticable due to lack of complete, reliable information about this person. People do not know the basic things about themselves. For example, they do not know which foods specifically they should eat, how the environment in which this or that person resides should be organized, in which occupations he can realize functioning implanted in him, which partners to choose to live with, to do business with, to reproductive with, and so on. All of this and much more one has to dream up, invent. But, what does this have to do with thinking?

It would be pertinent to compare modern *Homo sapiens* (in regard to which, from our point of view, it is not quite correct to use '*sapiens*' for reasons mentioned above) to a Neanderthal, who got his hands on a computer, and causes damage to this device by any of his actions because he does not know how to use it. A person does the same in respect to himself, his body and "Soul."

And, to continue the parallel, right under the nose of this Neanderthal is an instruction to the computer that he is holding in his hands, but he is so deep in his fantasies that he does not notice it. In addition, he does not know how to read. From our point of view, an analogous situation occurred with the "Catalog of Human Souls," which all this time was "under the nose" of humanity. Probably a Neanderthal too would have taken the instruction to the computer in his hands, then just tossed it, and continued to beat the unit that does not want to turn on with a stone ax. How else to explain that having the Book of Mountains and Seas (Shan Hai Jing has been translated into many languages) in global culture, for thousands of years no one could understand that it the Catalog of human population? From our point of view, there can be only one explanation: intellect of people with a dead "soul" does not work as it normally must work.

A person who is not familiar with the "Catalog of Human Souls" does not know and does not understand much of that what happens around him and within himself. In essence, in this state of unknowingness and lack of understanding he lives his entire life. He does not know why this or that event occurs in his life, why this or that person comes into his life, and so on. And, most importantly, it is unclear why he himself behaves in certain situations in one way and not in another, and why his behavior can differ greatly from algorithms of behavior of others. Not getting clear, specific, justified answers to these questions (in the best case, only other's interpretations, or fantasies), he replaces them with invented, made up explanations. Also, he does not know that all his explanations are invented by the "demon's head," and are not made on the basis of analysis by his natural intellect. He thinks that his explanations are made on the basis of the thinking process, but they are not; they are nothing but his fantasies, his imagination.

In search of recipes on how to live, first a human unsuccessfully tries to find a single clear system, scheme in culture. He does not find it, and goes into a mode of life known as "trial and error." He looks around and sees a variety of ways to live, many actions, solutions to various problems, and he has nothing left to do but to start trying one after another, hoping to find, as they say, "his own" philosophy, occupation, food, sex partner, passion, "Path," and so on. Although this has nothing to do with his personal, his natural—"his own." A person generally considers "his own" anything more or less comfortable from the point of view of work of psychophysiology, but more importantly, socially approved, socially meaningful, pleasant, comfortable, while it all has nothing to do with his natural program. (This is often called: "I understood it intuitively." Although intuition of a person with a dead "soul" does not work as a function—"the antenna is broken.") Choosing something from society, he tries to embed it in his psychophysiological existence. And, he does this with a greater or lesser degree of success—some part of that what he finds is used for his good, but most of the trials harm him (from slight damage to fatal consequences).

"Intellect" of a person with a dead "soul" searches and tests new and new options in regard to anything: food, partners, hobbies, occupation, exercise, cosmetics, treatments, physical development systems or systems of self-development. In this process, his short life passes by. The remainder of time left from search and endless trials he or she usually spends on liquidation of consequences of mistakes, although if the person had information about himself/herself from the "Catalog of Human Souls," he or she could spend it on something more interesting, useful and enjoyable.

This kind of life does not perturb a person with a dead "soul." Eventually, he gets used to the absence of positive results, and begins to perceive this as a normal course of his life.

Besides, regardless of results that a person with a dead "soul" gets, he is always convinced that anyone and anything is to blame, just not him. In his own view, he is always innocent, blameless, pure, sinless, and therefore cannot be wrong, mistaken. "Intellect" of a person with a dead "soul" was defined as narcissistic because the carrier thinks that he is absolutely wonderful, the best of the best, practically perfect. Although in reality the only thing in which "intellect" is really strong is ignorance, building of illusions, games, perversions and other abominations, which will be described below. However, such functioning makes a corpse out of a living subject.

It is also necessary to note that self-development, self-improvement and other terms that are "foolish" from the "demon's head" point of view, are usually not included in the scope of one's attention, his interests. His opinion on self-improvement is expressed well in the following Russian sayings: "The best is the enemy of good", "To teach someone who knows is only damaging", etc. He believes that since he is already like the Creator, why learn, why develop, why improve? And, the actual state of affairs is usually manifested in that he is not only far from perfection, but also immeasurably stupid, fits in with very few activities, does not know much and knows how to do very little. In other words, his real qualities are usually successfully camouflaged by the ability of the "demon's head" to build illusions, and this ability is trained for years. A "demon's head" is never too tired to self-improve in relation to any kind of perversion.

In the end, not wanting to know anything beyond the scope of fantasies created by his sick imagination, not wanting to accept reality as it is, not wanting to study and learn, not wanting to go towards anything, to move, to grow, to develop—a person with a dead "soul" remains dead. Although according to religious sources, as well as our research—human "soul," unlike his body, can be resurrected.

Sleep That Is Death

"...In appearance he is a human, but by the internal mood he is not a true human. As an untuned instrument that is this or that in its appearance, but it is not known by the sounds—this is how a person whose inside is torn by sin should be judged ... As the dead do not see, do not hear, do not move, a sinful person does not see, does not hear and does not move as a human being: he does deeds, but they are "dead." (Heb. 6:1; 9:14; Excerpt was taken from the Russian version.)

If you carefully observe those who based on quite a clear number of characteristics we consider to be people with a dead "soul," then the analogy between them and the dead becomes evident. Despite the fact that before us we see a seemingly alive someone who moves his body, eyes, head, arms, legs, opens his mouth to eat, drink, talk, goes somewhere, does something— we still agree on this issue with authors of religious texts that all this is only a parody of life. Since on the inside of this person is ringing emptiness. There is nothing on the inside because his "soul" has died.

Daily condition of a person with a dead "soul" can be compared to the state of a computer called "sleep mode." Although it is even better to compare it to biological anabiosis—one is alive, but all life processes are so slow that they are reduced to almost zero. His "Soul" (natural program) works at a maximum of two-three percent, only in order to support the physical existence of the body for some time.

Although not for long since "the battery" is charged only for a while. Once this natural energy charge that is given to a human from birth ends—his biological death will occur. A person with a dead "soul" has nowhere to recharge from, as the only "natural charge" for every living beings on this planet are natural images of his natural program. However, a person does not know anything about them, and does not study and use them. He lives on artificial images, which do not "charge." Artificial images are like those plastic apples that were mentioned above, they cannot be eaten to gain energy.

Precisely because of this, and not for any other reason, life of modern *Homo sapiens* is so short.

And, it is not only short, but also difficult. Since it is difficult for a person who is alive only by a few percent to not only support his existence, but also to bring up children, to keep contacts with other people, to each time try to fit in conditions that are offered by civilization, and to keep up with scientific and technological progress. Due to lack of access to information on natural images of his "Soul" from the "Catalog of Human Souls," the energy level of a person is so low that in response to the question "How are you?" one often hears: "Tired." Of course, if the person does not know you well, he will draw a smile on his face for a few seconds, and answer: "Fine, thank you;" although, in most cases, it is not possible to believe this after taking a close look at this person.

Due to the constant lack of energy to do all the things that a person is obligated to do through the day, he often gets severely irritated, and that, in one form or another, manifests to the world, to others, to himself. A person either falls into depression ("I am so sick of this damn life!") or he gets angry, and begins to destroy everything around him, including people: "I am so sick and tired of you all!" Or he begins to engage in semi-conscious or conscious self-destruction, according to the principle: "Life is shit, at least I will do something nice for myself." Although in fact, everything that a person does at such moments is break himself, himself as a natural mechanism.

A person with a dead "soul" is not truly happy about anything. And, even if something does please him, this joy is short-lived. And, moments of happiness are even more rare and much shorter. Whereas, according to information that we discovered in some of the ancient sources, happiness is a basic emotion, an emotion on the basis of which a human being must live to be healthy, powerful and to thrive. However, the basic emotional states of people with dead "souls" are anxiety, dispiritedness, irritation, hatred, and anger. These states corrode a person from the inside like sulfuric acid, and cause only one desire: revenge; to take revenge on the world, to befoul another person's life, to befoul his own life. And, as the worldwide psychiatric practice shows, no doctors, no medicine can help a person get out of this emotional state; the most they can do is "turn him off" together with these feelings and emotions, and turn him into a natural "vegetable" that feels nothing at all.

However, even without being under the influence of "help" of psychotropic drugs, a person does not want to do anything due to lack of sufficient energy and internal motivation. Hence, the occurrence of fantasies like "When I retire...", "When I earn a lot of money...", "When the kids grow up..." They imply that once that happens, he or she will not do anything. That is why people dream of only how they could relax. Sleep and food provide small energy charge, enough for a person with a dead "soul" to last for some time, but not for long. And so, he tries to somehow arrange his life in a way to reduce his level of activity (physical, psychical, intellectual, etc.) to almost zero. Often, one can hear people with a dead "soul" say: "I will do this some other time...", "This is too difficult for me to understand...", "Worrying is bad..." and so on. They hope to spend less vital energy that pours and pours, like sand in an hourglass, bringing the hour of their physical death closer. This necessary measure allows people with dead "souls" to live about 70-80 years. Although they live in a "turned off" state, not at full capacity, and, by and large, they do not really live. Life is something completely different.

On the other hand, by switching to the "economy mode," people with dead "souls" completely rule out the possibility of "becoming alive." It is as if they enclose themselves in a capsule. However, in order to "become alive," that is to revive his "Soul," to "turn on" his natural program, one must do the opposite: he needs to open himself to the world, to information from this world, to begin to desire, to seek, to study and learn new things, and to act, while making dreams and desires a reality. In other words: he needs to move.

People with a dead "soul" demonstrate a parody of movement—they go somewhere, run, and rush about. For some reason, they think that if they make movements at the physical level, sometimes purely mechanical ones, they can justifiably consider themselves alive, and then their lives will be full, interesting, bright, and eventful. And, that this will be life. From morning to night people make themselves do something; they invent a wide variety of things to do based on the following principle: "to keep myself busy." However, while performing some actions, they are not in reality, but in their "inner world" that their "virtual room" ("intellect") replaces for them. Although being in a "virtual room" as a space of one's imagination is very harmful. Especially if one does not live on the basis of his natural "operating system" because the "demon's head" that works on artificial images blocks natural desires, emotions, and aspirations, putting them into the category of "dreams." "Demon's head" offers to implement things that are not relevant to a person, things that he does not need, that are harmful, that do not provide anything to his psychophysiology—neither health, nor youth, or some interesting abilities, and do not bring him even a step closer to being a Human.

"Intellect" of a person with a dead "soul" immobilizes him on the inside, puts in a state of stupor, slowing down and stopping all internal processes of his psyche. That what they usually call by the words "calmness and grace" are calmness and grace of a graveyard, of the dead, and do not relate to the process of life; because that is lack of true movement, and primarily lack of motion of the "Soul." Panically afraid of everything that is new, afraid of changes, the "dead souls" put themselves in a capsule, and sometimes lose the keys to the "prison cell" forever. Due to this, their lives turn into a "Groundhog Day"—the same thing from day to day, from year to year, and from decade to decade. It is not surprising that boredom, then depression, and then—processes of self-destruction follow this.

The cause of depression is quite understandable: a person as if realized everything he wanted in his fantasies, but in reality he discovers that he did not do anything, did not move, and still lays as a corpse inside of another "Groundhog Day." In order for depression to not finish him off completely, a person is left to entertain the illusion that "such is life," that "all people live like this," and that his life is full of events on the grounds that today he tried a new cake or invented a new perversion.

Any animal has inner life; a primitive one, one that makes the internal processes a reality. If an animal wants to eat, or something else, it will not go to sleep in order to do it in its dreams, as from the point of view of nature this would be a perversion. While people live like this in 90% out of 100. If they want something, they go straight to the "theater," into the space of their fantasies and watch a "movie" about how they are doing it. Life of *Homo sapiens* is so automated that it is possible to dream even while driving a car, at work in front of a computer, and so on. However, once a person wakes up from his dreams, he feels nothing but frustration because a human experiences natural happiness as an emotion only when one of his natural (program) needs is satisfied in reality.

There is no better way to kill yourself than to live in your fantasies. Humanity only has yet to come up with devices that will eat, drink, go to the toilet, and sleep for them. Although this is unlikely, people do not stop wanting to fully move into the space of their dreams. They do not understand the threat. This is how artificial images work.

*****

From our point of view, a person with dead psyche is like a rotten tooth that stands for the time being, but realistically no longer belongs to the category of living.

*****

As our studies have shown, psyche ("Soul") is dead if a being of biological type *Homo sapiens*, by nature possessing incredible abilities, talents, and great potential knows nothing about it and does not use this. Metaphorically speaking, the treasures remain locked up in a room to which there is no key, no one can see them, make sure that they exist, look at them closely, let alone use them.

All psychophysiology of a subject with a dead "soul," including his intellect, is "idle." Natural program qualities, his functions do not get 100% realized. As a result, without having information from the "Catalog of Human Souls," figuratively speaking, a high-class Formula One bolide (that any person is by nature) lives his entire life as a primitive scooter. This "bolide" not only does not start up, it does not get cleaned, does not get repaired as it should, is not kept away from moisture and temperature changes, remains without care for many years in the open air and with a running engine. And, this continues up to the moment until corrosion and rot cover the body and parts get worn out. Then, the machine will finally stop for good.

The result of such "life" and "functioning" is that today a Human, a unique "Creation of God," is turned into a consumable material, which once it gets worn out and damaged (broken) will be sent to a garbage dump and decay. Is this what "God" created a human for? Are there not enough nutrients on planet Earth to feed all living without using human bodies as fertilizers? And, is a creature that lives this way really the "Image and Likeness" of "God?"

"...When a human, leaving all benefits, by disobedience ate the pernicious fruit, the name of this fruit is mortal sin, then he immediately died for a better life, exchanged the divine life for irrational and bestial..." We agree with this, however, in our opinion, there are no specifics; namely that the main "Mortal Sin" of a human is life that is not based on his natural program that was created for him by "God," as well as creation of artificial images.

"Affairs Of The Dead" And Affairs Of The Living. Resurrection Of "Soul" Is Possible.

As mentioned above, a person whose "soul" is alive differs from a person with a dead "soul" by lack of inner life, which his "Soul," his natural subtype program gives. When this program works, an individual has no problems with presence of motivations, aspirations, desires, as well as energy to implement all of this in reality; he does not have problems with existence of inner life. However, we suppose, it should be explained in more detail what exactly is meant by presence and absence of inner life. Certainly, in this format this will be done very briefly. (However, we review this in more detail in our other books on how a modern human lives on all six factors and what consequences this leads to.)

*****

One indicator of absence of inner life in a person is that he is only concerned with that what is outside of him: things, other people, events, and processes. His life is a never-ending stream of things related to the external world. And, he does not take care of his own self in terms of his natural program, his "Soul."

And, he himself does not exist, as an Individual—he disappears. To nature, he disappears as a valuable natural object as well. He is just a small unnoticeable screw of various societal processes.

And, as time passes, a person with a dead "soul" needs to be in the position of such a screw more and more. Fear of loneliness as a situation of facing the emptiness inside him, unknowing what to do in general and with his own self in particular are just some of the reasons. An even more important reason behind "cooking in a communal pot" is that since his psyche is dead, a person needs stimulus, motivation from the outside in order to move at least somewhere, to feel at least something, to want at least something, to strive for at least something. He does not have motives, desires inside him because his natural program does not work. A person with a dead "soul" is forced to seek both positive and negative stimuli in society.

And, his environment, the society willingly offers him a choice between fun entertainment and troubles, funny or scary, fun or dangerous. Civilization has a well organized and setup system of, so to speak, "stimulation of corpses," people with absence of their own desires, aspirations, internal experiences. Society offers many things, which they can entertain themselves with during another "Groundhog Day." A person is offered things that can please him, upset him, anger him, and so on. In short, that by which it is possible to feel, using the language of slang, the drive as a physiological desire, a need, a motivation, a stimulus, an internal impulse, excitation. That, what a person with a dead "soul" does not have, but also that what he needs in order to at least imitate life until his physical demise. Although in our opinion, all civilizational suggestions are quite primitive and boring, but that is our personal opinion.

Behind the seeming idyll (when an individual does not really need to do anything, does not need to think and want anything, as that is done for him, and he gets entertained) stands the fact that society is ready to stimulate and please him mostly in exchange for money. So there is another pastime for a person with a dead "soul": a job or a business. In the absence of understanding of what to do with himself a person goes to earn money. This, at least for some time, gets him out of the state of boredom and gives his life some meaning; although only for a while. People with dead "souls" sometimes enthusiastically engage in "money making." However, this does nothing except wear out their psychophysiology. And, it is not easier for those who do not have the need to engage in earning money because they have all that they need within an arm's reach, they have nothing to do, everything has already been tried, there are no goals ahead—these factors make the physical death of a person approach faster. Free or nearly free entertainment includes watching TV on the couch with a bottle of beer or vodka as an antidepressant, as well as "friendly chatter" with members of the household, colleagues, neighbors, in which one can recharge or emotionally drain, compete "who is cooler," get a lot of "good and wise" tips on how to live, and feel needed by someone (although this is not so since besides his acting in a role of a living confirmation that he is needed by another person, he is not interesting in any other way). However, this has nothing to do with friendship. All existing public formations in civilization fall under the category of stimulus and entertainment: from casinos, theaters and sporting events to courtrooms, senators' chambers, and hospitals. There is something for everyone. All this reminds of euthanasia because there are basically a million ways of how to throw one's life in the trash in an interesting, fun and pleasant manner, and never become a Human.

It is easy to guess the results of lack of one's internal life. From constant engagement with something or someone outside himself, his last strength evaporates, psyche gets exhausted, and physiology wears out. After "entertainment" offered by this civilization, a person gets into more and more prolonged states of depression, disappointment, and tiredness from life, up to suicidal thoughts and actions. On the other hand, without this civilizational "entertainment" he gets bored. And, boredom is not such an innocent thing because just out of boredom, people often do things that ruin their lives.

From this follows a conclusion that in people with a dead "soul" not only the intellect, but the emotional factor as well does not work correctly. Apart from the fact that a person feels nothing unless someone or something from the outside stimulates him, his constant companions are anxiety, fear, irritation, anger, and psychotropic drugs, but this was already mentioned above.

*****

Without having any idea about inner life, a person with a dead "soul" has no idea what it means to take care of his own self. He thinks that to "to take care of himself" means to be an egoist, to not come into contact with other people, to not take care of anyone else, to not take others into account, etc. This is very far from what these words mean to people who take care of their "Soul." That is, to those who know their natural program, study and take care of it.

To put it very briefly, the presence of inner life in a human assumes an entirely different functioning of a person. For example, his intellect is not occupied 99% of the time by "sitting in a theater;" in other words by fantasies, building of castles in the sky, self-consolation, and narcissism. Intellectual activity occurs under normal algorithms: a person can think, he has information (and if he does not, then he gathers it), analyzes it, and draws conclusions. Also, the function of imagination is used differently. However, most importantly, intellect is used for its intended purpose: to provide communication with the program, the "Soul" that he knows. For example, for collection and interpretation of information on images of his program, for monitoring of his actions with subsequent analysis, for setting new goals, and for solving problems related to himself, his own life. His intellect is not used only to come up with another artificial image, how he will live and present himself—he simply does not need to do it as he knows himself. Also, intellect is not involved in processing of the many informational sources constructed on artificial images, which are not possible to review, but more importantly, it is harmful and pointless because they do not carry any reliable information. Thus, a person releases a huge amount of time for other, more important, more useful, and more interesting activities. Also, his intellect is not involved in modeling perversions in relation to himself and to others (they will be described below), which beneficially reflects on all of his psychophysiology.

In the physical factor taking care of your own self implies active care in terms of health. This does not consist of swallowing food additives, vitamins and pills, or appointments with doctors. A person turns for help to the Catalog of Mountains and Seas for information on how images of his individual program live, and tries not to violate the natural laws in relation to lives of these images. In addition, the description of each program in the "Catalog of Human Souls" contains information about what exactly the right remedies are for a particular person. These people turn to doctors as to experts in the field of human physiology, but they consider their advice from the point of view that modern medicine has no knowledge of subtype differences within the biological type *Homo sapiens*. A mandatory element in everyday life of a person engaged in taking care of himself are sporting activities, but only those activities and only in those algorithms that are required by the natural program of the person. The range of these activities for any person is huge and it is not necessary to choose any single sport, as he can develop himself in all of them. The same can be said in regard to the wardrobe of a person. He knows exactly what materials, what styles, shapes and colors of clothing, shoes and accessories to wear based on his program's particularities. The range is also wide. The same can be said about jewelry, if it exists in the individual program of a person—he does not need to wear something on his body that is not related to him personally. The same applies to tattoos, if it exists in the individual program of a person, he does not need to put on his body artificial images that will be killing him; he knows his natural images and in the form of tattoos on his body they will bring him only benefits.

In the nutritional (dietological) factor a person will never get food from "a communal pot," regardless of whether it is a fast food establishment or a posh restaurant. Since he knows that there the same dishes are prepared for all, and that is unacceptable because all people are representatives of different subtypes. And, naturally each subtype has its nutritional range, going beyond which is a health risk. Especially in the case of humans; each subtype has a wide range of what can be eaten, but each subtype has its own range. Therefore, with the style of nutrition that was called "eating from a communal pot," a person puts into himself, figuratively speaking, as into a car, anything instead of proper grades of gasoline: from milk to hydrochloric acid. As a result, if there is a financial possibility, a person who knows his natural program from the "Catalog of Human Souls" hires a good professional chef, who cooks for him personally. Or, he prepares dishes himself since he knows that food is more important and better than any medicine.

And so on in relation to all remaining factors: emotional, sexual, and environmental.

Taking care of yourself, your natural program is so interesting, engaging, pleasant, and exciting that days fly by. There is no boredom or need for "entertainment," which were mentioned above. An individual finally becomes interesting to himself. Finally, he knows what to do during the day, month, year, years. This frees him from pathological dependence on anyone, dependence both from individuals and from society as a whole. A person becomes self-sufficient, autonomous.

However, this does not mean that he does not communicate with anyone. On the contrary, his circle of communication expands greatly. However, a person always has the option to decide whether it is necessary for him to communicate with someone at this or that time or not. He can no longer be "pulled like a carrot" by someone for some purpose. A person knows why, for what purpose he will communicate, and what he primarily wants from this or that contact. And, he knows how to get it. And, he knows how to do it right, so that this communication is constructive, effective and pleasant for both his opponents and for him. Thus, the society begins to get used for its intended purpose, namely to take that what a person needs from it. And, in order for a person to get that what he wants from another person, he no longer needs to deal with those abominations with which people unfamiliar with the "Catalog of Human Souls" usually reach their goals—he reaches his goals by dignified methods.

However, more importantly, taking care of yourself pays off in the form of appearance of new qualities in a person: skills and abilities, as well as psychical and physical well-being. And, it does not matter what a person who takes care of himself engages in: household chores, work, communication, studies, work on his body, etc. Any of these processes take place on the basis of his internal motions—natural needs, desires, goals, and objectives. His "Soul" is alive, it works, and he is not empty inside. Unlike a person with a dead "soul," he is not an empty cardboard box.

Sooner or later, even with seemingly small changes in his appearance (although usually external changes do occur), a person who has an inner life, takes care of his "Soul," begins to see changes in his personal qualities and augmentation on all factors. Another important prize for pleasant, but still daily work directed at activation and maintenance of his inner life is that he finally stops being a person with a dead "soul." He gets filled with life on the inside. He begins to not only desire, dream, but to make real actions in order to realize all this. He begins to create, and create in the vector of "life," not death. Though he makes mistakes, and not everything turns out as it should right away—it is okay. Most importantly, a person starts to move in the broadest sense of the word. And, this movement begins to move him to new heights.

And, it might seem that nothing is happening with this individual, but it just seems that way because many processes, a number of events occur on the inside. In contrast to a dead person fixed "in a pose of a monument," a person who is engaged with his natural program from the "Catalog of Human Souls" moves forward, grows, and creates. Unlike life of a "dreaming corpse," this allows him to get real results, to demonstrate real achievements that can be seen, heard, and touched. It might be some interesting creative product, change in his social status, power, financial situation—anything. He gets all of this instead of castles in the sky of a person with a dead "soul," who does have all of this, but for the most part only in his dreams.

Another very important result of a person's movement from death towards life is an increase in his cultural level, and together with that, the level of his survival in the environment. After all, culture is the basis of survival in the environment for a single individual or a social group. And, what culture could a person with a dead "soul" (who aimlessly, ineptly spends his life and gets constantly used outside of the range of his natural program) have? The most he can do in an attempt to increase his cultural level is to fill his brain with a pile of unreliable and indigestible nonsense that has nothing to do with reality since it was established on the basis of artificial images. Low cultural level of an individual, including lack of information is the cause of lower survival rates of people with dead "souls." Their lives are even shorter than of some animals, for example, red sea urchins, bowhead whales or turtles that live for 200 years or longer.

"...Want to be healthy? Want to be saved? If you do, My wisdom will guide you, My mercy will have mercy on you, and My might will help you and save you. If you do not want to be saved, if you want to run from the Eternal Life, if you love your death more than salvation, neither My wisdom nor My mercy nor My might will help you. Can warm wax adhere to ice? It cannot! And My mercy, My wisdom and all My might cannot adhere to you if your heart is cold as ice, and has no warmth of desire to be saved. When you do want to be saved, I will be happy to help you..." (Dimitry of Rostov) The "Catalog of Human Souls" as a real source of information left by the Creator to humanity—exists. And, it has information for every person who wants to resurrect his "soul," to become alive again. The only requirement is desire, your own decision. The Creator offers help to anyone at any time, but people do not hear His voice. And, they do not hear the voice of their "Soul." "Demon's head" that replaced a person's intellect, constantly mutters something into his ear: interprets events and behaviors of others, his own behavior, what he should do, what he should be, how he should act, what he should strive for. A person unfamiliar with his own "Soul," naively mistakes this voice for the voice of his own mind, for the voice of his own "Soul." However, in reality that what speaks to him is a soulless artificial "operating system."

Sleep Of The "Soul," Not The Mind Produces Monsters. Cadaveric Stains Of Sins.

"...As worms breed in a dead body, in a soul, which lost Divine grace, bred like worms: envy, deceit, lies, hatred, enmity, fighting, rancor, slander, anger, rage, sadness, vanity, revenge, pride, arrogance, unmercifulness, extortion, theft, falsehood, irrational desire, gossip, slander, ridicule, renounship, perjury, curse forgetfulness, insolence, impudence, and every other evil, hateful to "God"; so that a human has ceased to be the image and likeness of "God," as he was first created, and began to be the image and likeness of the devil, from which all evil comes..." (St. Symeon the New Theologian (60,45); Excerpt was taken from the Russian version.)

As mentioned above, someone who by his own volition or due to unawareness does not use his psyche—becomes a living corpse, a person with a dead "soul." And, the appearance of such a person is very deceptive.

A corpse too looks fresh at first, even though it is already dead. The same is easily observable with people-mutants. In their youth, they also look fresh, sometimes even very attractive. However, in adulthood any one of them becomes increasingly, so to speak, "fragrant." Outside attractiveness melts more and more, the "aroma" increases, and at an old age a person turns into an unattractive and very difficult to stand "toxic waste." Since any corpse, by definition is no longer capable of anything except decomposing and "delighting" by its miasmas. Therefore, it should be noted that a person with a dead "soul" engages with the world around him in a very peculiar way: anything that he or she touches—he damages with a varying degree of severity, thereby destroying all living and nonliving things around his. This is the process that we mean by "aromatization."

"Aromatization of space" is spreading entropy around yourself. By entropy we mean chaos brought into the lives of others, frank wrecking in any form, destruction, demolition. The tools used to make damage are described in detail in the ancient book that we studied. Here are some examples: to act cunningly, to do something secretly, to seek by illegal means; desperate falseness, shameless lies, trickery; libel, slander; peaching; to instigate, to incite; to banter secretly; to gain favor out of mercenary motives; gloomy (depressing, disturbing) atmosphere; atmosphere of a mystery; a secret, a mystery; "devil's tricks" (stunts, secret plans, hidden intentions); demonic affairs, a mischief; to resort to tricks, to act dishonestly, to act in secret; to act furtively, surreptitiously, secretly; and much more from the same category. From our point of view, it does not make sense to comment on the above list, as each person living in this civilization faces all this abomination daily, and not only outside, but inside himself as well. It should also be noted that seemingly harmless expressions of people with dead "souls" (their, supposedly, jokes, mistakes, games, lies, embellishment, meaningless actions, aimless behavior) are always far from harmless consequences in the form of destabilization, chaos, and destruction, introduced by them into the environment, in which they live. This is the "aromatization of space."

For example, consider speech of a person with dead psyche. This speech is a reflection of his pathological consciousness, pathological perception of the world, and therefore it is insincere, deceitful falseness. After all, ever since childhood they have been taught to wear masks, to play roles. Speech as a means of communication reflects the basic principles on which communication is built. This principle is lying. It got to the point that excuses and beautiful names were invented for lies: "a white lie," etc. However, people who live on the principle of lying often do not know that a lie, falseness is nothing but a tool of murder.

And, people-"evil spirits" widely use this tool for destruction, as they do not know how to do anything else. Indeed, if we look at any of their manifestations directed outward (fraud, deceit, gossip, peaching, libel, slander, intrigues), there is always damage to the person against whom they are used. The only question is, to what extent: as a scratch or a knife stab. In any case, sometimes people die even from scratches, for example, if there is any infection. And, what if a person is covered with minor scratches from head to toe, all over his body? More often than not this is what people do in relation to each other.

One might think: what is so terrible in a lie, how can it seriously damage a person? However, the ancient sources of knowledge state that through lies a systematic killing of life occurs. And, they clearly speak of the fact that if you lie, you are killing life in your own self. A lie, however decorated, is poison, and it is similar to eating broken pieces of glass—a microscopic dose, but at each meal. According to the laws of nature, death of the receiver will occur sooner than if pieces of glass were not added to food.

Lies or falsehood, being the basis of this civilization, bring millions of people to the grave daily. It is not necessary to believe this; continue to water yourself like a flower with a weak solution of hydrochloric acid, only then there will be no way to learn the results. Since nature is silent, it does not live in the world of words, it lives in the world of real actions. Do not expect it to warn and persuade, one day it will simply sum up human actions.

To "Be Yourself" Means To Be A Corpse?

All people with dead "Souls" wear masks and almost never take them off. One can be "himself" only when he is alone, completely alone. Although when people with dead "souls" are left alone they usually get nervous very quickly; they are so terrified of their internal "content" that they immediately begin to look for at least someone just to stop looking deep into this frightening abyss.

Places where a person with a dead "soul" does not have to put on frills and show off are a desert island and a bathroom, the door to which closes. The older people get, the more they do not understand the meaning of the words "to be yourself." Masks firmly adhere to their faces. However, the civilization is constructed in such a way that without the "Catalog of Human Souls," without knowledge of which qualities and properties of an individual hide behind this or that physical form, it imposes certain standards of behavior on people. One who is not ready to become an outcast (and, no person with a dead "soul" is ready to become an outcast because, as described above, he depends on society) must play social roles throughout his entire life; roles that have no relation to his "Soul," his essence, to his natural program, to how "God" created him. So people are made recognizable, and their behavior predictable and manageable. This is very easy to implement in practice. Secular ideologists invent and launch into use a certain number of artificial images, and offer to choose according to your taste. People are only left to choose from that what is offered, and "get into the skin" of selected images that will make them as it is necessary. This is how any image functions.

People accrete to artificial means (usually, it is not one, but many, meaning a typical schizophrenia) that the secular civilization imposed on them so much that their natural individuality disappears completely. Playing with images cannot be done by anyone, not even by ideologists. A person only thinks that he will remove the mask that he is wearing now. However, that is not the case because an image works in a different way—a person becomes a carrier of qualities and functions of an artificial image, fully. Although he does not even notice this being absorbed in that he fooled everyone by pretending to be someone he is not; in reality—he got fooled, and fooled greatly.

However, as already mentioned, the natural program of a person works anyway, at least a few percent of it does. Therefore, any artificial image, using the computer terms, "installed" next to the real program conflicts with it. As it turned out, this conflict is the main cause of occurrence of any disease in humans, and first of all psychical disorders. Coexistence of a natural program in an individual that supports his physiological existence with artificial images, comes with heavy losses in terms of health, and takes a huge amount of energy. And, lycanthropy pumps out the last strength out of a human. To create and maintain his "image," or rather numerous "images", each time a person must stretch all his psychophysiology to the limit, deforming his body and psyche. It is not surprising that even in countries with a high quality of life many people have not just serious somatic disorders, but are crazy according to psychiatric criteria.

Lycanthropy means to not live your own life, and to have no idea about what it is. "Living your own life" from the standpoint of artificial "operating system" means anything except what it actually means; for example: hypertrophied egocentrism, strangeness, reclusion, etc. It is impossible to live your own life and cram yourself into social roles as into artificial images. Life of a person with a dead "soul" is an ongoing performance, a show, self-presentation, where there is no one behind the mask, the facade. Although this seems harmless ("No big deal, it is just a game.") such way of life is a powerful tool for destruction of body, "Soul," and Individuality. In order to correspond to images that by nature you are not, you need to destroy your Individuality. The result is frozen masks everywhere, and the people behind them are dead.

A heavy burden of a person with a dead "soul" is constant need to hide something, to conceal, to do everything so that no one ever guesses what he or she is really like. Since otherwise the beautifully painted pastoral will "go down the drain." Besides, every day people have to lie to each other that they are alive. This takes away energy as well. As a result, as soon as a person is left alone, he immediately deflates like a balloon, and starting this moment presents a tired, confused, disoriented, weak, helpless, sick, miserable, insignificant, and pitiful creature. This is what people with dead "souls" mean by "being themselves"—being a natural corpse that does not need to pretend to be a zinger under the slogan "better than others."

Not "Human And Nature," But "Nature And Human." An Elephant And A Pug.

Another characteristic of people with dead "souls" is that they try to fight Nature, to do things contrary to Nature by any means and at any cost. This tendency has been present in this civilization probably since the disappearance of the "Catalog of Human Souls," meaning for many centuries. By nature we mean the external natural environment in which people-mutants live, as well as their inner nature—psyche and body. They seek to win over both with persistence of a kamikaze. By the way, the toolkit to fight nature within your own self is the same: lies, cheating, devious plans, tricks, etc.

As far as nature that is outside of people with dead "souls," it does not matter if they are trying to turn the rivers back or make another monster genetically. The most important thing for them is to say: "I won over nature!" Trying to escape from reality (and nature in particular) in their illusory "house of cards," people-mutants are literally obsessed by the idea of subordinating the surrounding nature to their sick intellect, although this victory is impossible. People-mutants, in a state of enthusiastic fighting with nature, are not bothered even by the prospect of not being able to celebrate victory because they are part of nature and will die with it. Thank "God," this is impossible.

Intellect of mutated people is so underdeveloped and functions so poorly that, despite their own research, including of cosmos, which has shown long ago the power and scope of their "enemy," they still have not given up on their delusional ideas. They also have not been affected by the experience of ancestors, who considered themselves great dodgers and brilliant schemers, but have passed away, while nature continues to exist: the earth spins as is spun in one direction, the sun continues to shine like it did before. And, a person with a dead "soul" has never won over neither the earth nor the sun, not even his own cat. And, all attempts to cheat, to fight their own nature always end only in death because they are nature and by trying to kill it, they kill themselves. Only, external nature is so powerful as a mechanism that it does not notice attacks by people-mutants.

Humanity should ponder their true scale in relation to nature. After all, nature is not planet Earth or even the Solar system; the Solar system on the scale of nature is a tiny part, a molecule, an atom, a cell inside something more grandiose. Would it not be more reasonable to work towards reaching maximum synchronization of your own self with your own nature and nature in general? Especially since every human has inbuilt potential of a being capable of becoming as omnipotent, powerful, and on the same scale as nature. Maybe it would make sense to stop being a pug that barks at the elephant, and start to learn from nature?

Genius Perverts

So far people prefer to consider themselves great and brilliant, without thinking that in the case of people who live based on artificial images this is genius of the perverts. Considering himself a genius is another distinguishing feature of a person with a dead "soul." It is not surprising that their genius never goes out of the range that was called "cadaveric stains." Geniuses in this civilization are devious, deft scammers, mythmakers of all types, possessed, various types off wreckers, and backbiters.

One can observe brilliant cheaters, gamblers, drunkards, debauchees, scientists, musicians, writers, businessmen, etc. everywhere. Despite the fact that in society they are clearly divided into "good" and "bad," in the ancient sources they are combined into a single group called "genius perverts." From the standpoint of these sources, if we are talking about a person with a dead "soul" (someone who lives and functions on the basis of artificial images), then there is no difference between him and a genius criminal, a murderer, or a swindler and a genius scientist, who created a crazy theory that misled humanity for many years and put a break on all science. Neither one nor the other can be classified in any other way than perversion and wrecking.

Any person with a dead "soul" is a pervert. However, there are mediocre perverts and perverts of the highest standard. The first ones brilliantly harm only themselves and their immediate surroundings. For example, people who are sick, but refuse to take medicine and do procedures that can cure them; or those who need money, but refuse to make it; or ladies, who run around with navels bared, in heels and stockings in freezing cold; or showmen, who do devil knows what, including with their own selves, just for the sake of entertaining the crowd and gaining fame "for an hour"; there is a huge number of other examples. However, these "geniuses" are not particularly outstanding. A different matter is to, for example, make a statement about unknowability of the subconscious in principle, or that a human evolved from an ape; in other words, to create a theory, which for many years will significantly slow down the development of scientific thought. All brilliant perverts imprinted in human history differed from other "geniuses" exclusively by the scale of their perversions and the degree of harm caused by their activities. They can be called the genius dead because consequences of their wrecking are large-scale and impressive as far as the degree of harm inflicted on people and human culture.

From the standpoint of the ancient sources, scientific and technical progress (closely related to "genius" on the basis of artificial images) is a litmus paper that shows low (in terms of human resources) quality of the level of development of civilization and extreme helplessness of humanity. Consider Jesus Christ who, according to the story, insisted that he was not "God," but a Human, who walked on water as on land, who turned water into wine, who resurrected the dead, and replicated food—it is unlikely that he would need modern means of communication and transportation, medicine or supermarkets. Scientific and technological progress achievements instead of processes aimed at transforming the human resource to at least the level of Christ, in fact, are not achievements, but a complete failure. Humanity would not need medicine, would not need any kind of modern means of transportation and communication because by nature any person has abilities to communicate telepathically.

There would not be a need for credit, financial, banking institutions; supermarkets; construction companies; judicial system; a huge number of diverse industries because just as easily as Jesus turned water into wine, any person could turn anything into money, gold, gasoline, kerosene, penicillin, food, clothing, matches, a vacuum cleaner, a washing machine, and so on. A person who among his abilities has a possibility to exist in any environment, does not need anything of that what today's civilization offers.

Scientific and technological progress is not able to help a human being become a Human, like the "Son of God." Moreover, due to genius slaves of their vices, perversions and their self-fertile intellect, there are no possibilities of appearance of people with qualities of Jesus Christ. As soon as genius "living corpses" of all kinds, at all levels, from all social circles only hear about this, they immediately attribute it to the category of another utopia, while knowing absolutely nothing about the existence of the "Catalog of Human Souls," true human potential and true human destiny;

not even about who a human being is and how he functions. However, they do not require this knowledge because in their mad fantasies they are geniuses that are something. And, they are not bothered that they are leading humanity in a direction opposite of life, opposite of emergence of a new civilization, where people with abilities that greatly exceed any technical "miracles" are born and live. Geniuses are zombies who execute an artificial program provided to them.

"Ye Shall Know Them By Their Fruits"

"Ye shall know them by their fruits. Do men gather grapes of thorns, or figs of thistles?" (Matt. 7:16)

Now, regarding opinions about mutations on the basis of artificial images. One can feel whatever he wants about information that biological type *Homo sapiens* mutates without the "Catalog of Human Souls." One can feel whatever he wants about us, researchers who discovered this phenomenon: disagree with us, hate us because we are so well aware of the details of this "kitchen," and openly share "secrets of the mutants." However, that does not change and will not change anything.

For many centuries humanity prefers to amuse themselves by the thought that what cannot be seen does not exist. It was like that, for example, with the phenomenon called "soul," which, as it turned out, does exist. However, facts of real lives of people are available for observation, and cannot be attributed to invisible phenomena. At least by those whose sensory organs (sight, hearing, and so on) function, and who can analyze information.

In view of these facts, something unimaginable is happening in the human community: from ridiculous, inexplicable (from the standpoint of common sense and logic) oddities and unconformities, to severe perversions. Here are some examples from the first category. People prefer food cooked by someone, somewhere, in unknown conditions to food cooked in their own kitchen; to natural cotton, wool, silk, leather people prefer synthetic materials; instead of porcelain, faience, glass, and silver dishes the mass uses disposable plates; houses are not built from natural wood, stone, etc., and people calmly live in cardboard homes and believe that this is the norm; having the ability to place natural plants and flowers indoors, people use artificial ones; knowing that nature is a resource of any kind of remedies, cosmetics, items of interior, aromatizers, etc. people create artificial analogs, and use completely ridiculous reasons to explain this.

What else if not strangeness can be called the act of purchasing "organic" fruits and vegetables that are three times more expensive, while not giving up consumption of "fast food" at lunchtime, clothing made of synthetic materials (even while exercising), chemicals in the form of cosmetics and medicine, mattresses, pillows and blankets (on which a person spends about eight hours daily) packed with synthetics instead of feather, down, and other materials that feather and down can be replaced with? Tobacco, sugar, salt, caffeine, essential oils, sun rays are attributes to the category of harmful, while detergents containing chlorine are used virtually every day: people wash dishes, do laundry, clean surfaces in their homes, etc. By what if not, to put it mildly, strangeness can be characterized the fact that after eating, for example, artificial food coloring (currently, most dyes used are synthetic), covering skin by polyester, capron, polyethylene, nylon, polyacrylic, and breathing chlorine people genuinely wonder why they get diseases of the gastrointestinal tract, skin problems, etc. Whereas even a dog spoiled by home "nurturing" and conditions, would never prefer his "favorite" multi-component Pedigree to a plain, "single-component" beef bone.

"...Ye have heard that it was said by them of old time: Thou shalt not kill; and whosoever shall kill shall be in danger of the judgment..."

As far as more serious consequences of visible manifestations of mutations of *Homo sapiens*, here is an example of a phenomenon that according to both religious commandments and laws of any country in the world is an extraordinary violation. This refers to murder. However, in the community of people-mutants murder is commonplace, routine. Murders are everywhere. From our point of view, the situation is well illustrated by a quote from the Inspector Rex series: "As a friend of mine, a forensic medical examiner, said: "If candles burned on the graves of the dead, night would be as bright as day at the cemetery." People kill themselves, parents kill their children, children kill parents, representatives of one social group kill members of another social group, etc. However, if it is normal to kill yourself, then survival as an inborn instinct and form of behavior does not exist in all living beings? Is suicide in vain considered a violation of functioning of natural processes of life? If children do not kill their parents, then a child drug addict, or any "inapt child" (in this civilization every child is inapt) by his way of life and behavior strengthens health of his or her parents, improves their mood, and prolongs life? If parents do not kill their children, then how to classify the behavior of parents, who try to raise and educate a creature they know absolutely nothing about; instead of acting wisely and to the benefit of a child, they force him to live in a way that is unnatural to him, only because they think it is correct, useful, they want it and like it, and it is comfortable to them?

And, it does not matter that the civilization does not know the fact that in the case of *Homo sapiens*, children are not direct continuations of their parents, like an antenna; a representative of one subtype has children who belong to other, different subtypes. The consequences do not depend on knowledge or unawareness in this case. (It is not possible to explain in this overview what unawareness of this is fraught with. However, this is described in detail in our book titled Ahnenerbe—Your Killer Is Under Your Skin.) We will not continue on with the list of examples that prove that people engage in murder to relation to each other. We will just say that according to our research this is so, and we are ready to confirm this with facts.

At the same time, each person believes that someone else engages in murder, kills, but only not he himself, personally. However, the following situation that causes direct and deliberate damage to another person is familiar to absolutely everyone: when one person is asked to do something by another person, he replies "Yes, of course," then he does exactly the opposite, and explains his behavior to himself in any way from "I would be happy to fulfill his (her) request, but circumstances did not allow," to "You're dreaming that I'm going to go and fulfill your wishes, you think you found an idiot?!" Another scheme that is common everywhere is: "Let's establish close contact so that I could hurt you." It is probably unnecessary to explain how this happens, as each representative of this civilization has done similar in his or her life. For those who have forgotten, in short, this refers to when at the initial stage of a relationship, one person does everything in communication with another person in order for that other person to feel well and comfortable, but at some point he stops doing it, and begins to wreck in one form or another and in various amounts. This reminds of the following principle: "It is desirable to be as close as possible to your victim." In hand-to-hand combat this rule is as follows: "Use any "lull" in order to get closer and explode in the most unexpected places, break the psyche, bringing pain and destruction." In society this is called by words "friendship", "close relationships", etc.

From our point of view, conscious infliction of any damage is attempted murder, as any damage brings a human closer to death and any damage could potentially be "the last drop" for him. In general, if such behavior, and all the facts listed above do not relate to mutations as deviations from functioning of a person with a normal psyche, then how else can this be classified?

PART 11

BARBIE DOLL AND SUPERMAN—
SODOM AND GOMORRAH OF TODAY

Science can continue to try to explain the beginning of Apocalypse as natural disasters that are not connected with qualities of *Homo sapiens* and their way of life, with the level of development of human civilization. And, *Homo sapiens* can declare any number of concerns about pollution of the environment, even though nature has demonstrated that it can deal with it; that it is not the toxic waste from their activities that perturbs and tires mother-nature, but they themselves; that people with dead "souls," people-mutants are what nature regards as really toxic waste to be destroyed; that it is specifically because in their community there is such a phenomenon as "natural death" as death from diseases and old age; that for this reason, nature unleashes its righteous fury on humanity in the form of some natural phenomenon more and more often. In fact, nature, like a good hostess, maintains cleanliness and order in the house: cleans, washes, sweeps, etc.

Nature, which was also created by our Creator, can be understood: Shan Hai Jing, as the "Catalog of Human Souls" was right under the nose of humanity for many thousands of years. Just take the book off a shelf, no one is hiding it from you, read it, try to understand what is written in it, and once you do—make yourself in accordance with the "Divine Blueprint." Instead, a person says: "No, this book is too complicated for me, I'd rather watch TV. Let them entertain me, as I am great and brilliant." Or: "Why should I study nature? Let the botanists study it", "Why do I need to study myself? I already know everything about myself. And, if I get sick I will go to doctors or psychologists," etc. At the same time, a person's knowledge of himself consists of, figuratively speaking, only that he has two ears, and only because they can be seen in a mirror.

Nature, including his own, does not interest a person-mutant. He believes that he does not need nature. He or she says: "I want to be like Superman (Barbie Doll, Wonder woman, Spiderman, some actor or actress, Batman, Elektra, Daredevil, DJ Doe, Catwoman, Scarlett O'Hara or Natasha Rostova, Zorro or palled Macho Hulk, Gaius Julius Caesar, Al Capone, "femme fatale," and so on)." That is, instead of studying their real natural selves, stupefied and lazed humanity goes into mass culture to get another artificial image to transform into. Any "sheep Dolly" simply fades before the experiments of this level and scale. Since, as we have already reported, according to not only our research, any image works at the cellular level.

Based on the above, by the level of brutal consequences, none of the most perverted, crazy, cruel, horrible genetic, virological, Gestapo or any other experiments on a human can be compared with that what ordinary people do to themselves every day, and, notably, absolutely voluntary.

By severity, no sadist "of science" can compare to an average person, who executes terrible experiments on himself. Regardless of whether he is German, French, Russian, American, Chinese, Arab, Latino, etc. since mutations that arise on the basis of artificial images is a general civilizational phenomenon. Famous "doctor" Josef Mengele, "Angel of Death from Auschwitz" (victims of his experiments with human psychophysiology were dozens or even hundreds of thousands of people), is just a kid in his level of brutality in comparison with any man or woman of this civilization. Since any man and any woman in this civilization mock themselves, their own nature methodically and daily, at their own will, based on their own decision.

Society is forced to keep up with individuals. Although in, for example, American culture, an image of a superhero can be an ordinary human as well as a mutant, but then no one knows what to do with brood of mutant-monsters, who are no longer interested in anything except perversions and vices. To solve this problem, the society is ready to invent even AIDS, even an electric chair, even another "medicine" that will kill half a nation, only to somehow minimize it. After all, no calls to respect the religion, laws, traditions, history, and no "correct standards of life" can make people out of monsters.

*****

From our point of view, humanity has been playing with their toys too much: politics, economy, scientific and technological progress, terrorism, war and counteractions with each other, ideology and philosophy, the "golden billions," etc. While playing, humanity did not notice how some very adverse, to put it mildly, changes, in other words—mutations, happened to them. However, if playing continues, we think it is possible to miss the moment of the end of this civilization on the physical level. After all, natural mechanisms can neither be cheated nor persuaded—one not so wonderful day for this humanity, they will simply bring the process of destruction of this civilization to a logical end. Lives of different people illustrate this natural mechanism very well: no matter what you fantasize about, no matter who you present yourself as, no matter what games you play—mother-nature will destroy you sooner or later anyway. And, it will dispose of you only for one reason: you are a mutant. You are breaking the laws of nature (the moral and ethical laws of this civilization to not even need not be mentioned), the laws created by the Creator because nature is a product of the Creator's activity.

Residents of Sodom and Gomorrah (the fact of their destruction is described in the Bible) probably considered themselves great as well, and their activities the most correct, reasonable and purposeful.

According to words of researchers, "God's Vengeance" could be done in the form of a natural disaster, which will erase all humanity from the face of the Earth in one moment. In light of the above, this sounds, alas, quite logical and reasonable. Instead of methodically destroying each person one by one, it is easier for the Creator to destroy this whole civilization, wipe it off the face of the planet in a single moment, and create a new generation of people on Earth.

Humanity has the right to believe in this or not, but it does not affect the work of natural mechanisms of utilization. Also, according to information that we have, lack of knowledge about existence of the "Catalog of Human Souls" does not save one from the "Lord's Punishment."

How important it is to follow the realization of the "Divine Plan" (or, using the scientific language, knowledge of laws of nature, and following of these laws) is more than convincingly described in the story about the Deluge. One does not have to believe it, but the facts speak for that humanity has already paid a high price for attempts of people-mutants to fight Nature. The Apocalypse has already begun: natural and technogenic disasters, madness everywhere, drugs, rape, diseases, and death in massive numbers. But, apparently, the skeptics are right—this is not the limit yet. Also, let's not forget about space, from where some asteroid or another Nibiru can fly at any moment. Humanity cannot prevent, and even more so stand against a tsunami or a tornado, and so in this case one should not fantasize about salvation. No Superman or Batman will save anyone either, at least because they do not exist, they are fictional characters, artificial images.

PART 12

"PARADISE" IS HERE. "GOD" IS NEAR.

According to facts observable in reality, humanity is in great danger, and it is not without reason that both secular researchers and representatives of religious confessions persistently talk about the End of the World. However, nowhere does it say that this terrible prospect is the result of mutations of the biological type *Homo s.*, and that this can be avoided if people stop creating artificial images for their "operating systems," and instead live and function on the basis of natural images of their programs, given to them from birth.

It is not worth it to once again entertain illusions that nature requires humanity as a biological type. Objective facts speak for that nature is created in such a way that it can exist without this biological type. If humanity will disappear from the face of the earth at some point, everything else on earth will continue to exist, pant, work, flourish. By the way, if one looks at the Earth with the help of NASA technology, it seems uninhabited—people are not seen, as if they are not there. Therefore, it is likely that it makes sense for humanity to think about their true scope and role in the process of existence of this planet. And, maybe try to do something so that they continue to exist on this wonderful planet. Especially since it became clear that neither second Jesus, nor Batman will help humanity.

*****

Indeed, our planet is a truly amazing Creation, a finest and flawless work that amazes by the level of professionalism and grandiosity. According to some of the information that we got, humanity in vain invented a heap of preposterous fairytales about "Paradise" and "Hell" because this prevents people from noticing that the planet Earth is "Paradise." "Paradise" as an ideal place that "God" created for a human.

From our point of view, the Creator has envisaged and embodied absolutely everything so that *Homo sapiens* were healthy, wealthy (not only in terms of financial abilities) and happy. By its design and functioning, planet Earth reminds of a well arranged, organized at the highest level, elite kindergarten for *Homo sapiens*—live, grow, learn and become a Human. From our point of view, people do not notice this fact just because so far they have very little knowledge about the natural arrangement of the world that they live in (the external world as well as their own, internal). And, also because of the fact that artificial images ("evil demons," created no matter by whom: people themselves or someone else) torture, torment, tyrannize people on the inside so much that the world does not seem very nice to them at all. They see "Hell" around them instead of seeing "Paradise." Although "Hell" is actually that what currently happens inside of them.

Our Creator approached the creation of an ideal environment for *Homo sapiens* so thoughtfully and solicitously that on this planet there is everything that is required to support human life: food, water, shelter, clothing, remedies, etc. And, the book with descriptions of "Soul" structures of humans was left as well, as a source of information about what was left by "Our Father" for whom, what for and how to use these natural treasures; and, most importantly, how to directly communicate with Him through this information.

Another interesting piece of information that we have learned from the ancient books is that the Creator, so to speak, created humanity from his own self. We will explain. According to the data that we discovered in one of the sources, the Creator is so powerful because his "operating system" includes absolutely all natural images that are present in human programs. The entire spectrum, the entire range. Based on this, it is possible to suggest that His name is Human. Human as a being, which through processes of growth and learning on the basis of natural images, grew from the level of development of modern humanity to the highest level of development. The basis for the above stated are ancient texts. However, the Creator created natural images for our "souls," our natural "operating system" not like people create artificial images. A human creates artificial images in the space of his "virtual room" that is related to intellect (to the brain). The Creator took images for his Creation from his own self. Apparently, that is what was meant in the biblical "Let us make humans in Our Image, in Our Likeness." "God" created humans from his own self, and this is why knowledge, study of one's own natural images as a process of establishing connection with his own inner nature, with his psychophysiology is communication with "God," a connection to Him. According to information from sources that we study, connection with "Our Father" is done through the channels of natural images of human programs.

Based on stated above, in our opinion, humanity should stop worrying about lost "Paradise" and lost connection with "God" because now everyone has the opportunity to restore the lost channel of communication with the "Father."

*****

We are involved exclusively with science, and have nothing to do with religion. However, we assume that it is not secular social, but religious institutions that should take the mission of saving humanity from destruction because it is religion that originally deals with an object called "human soul."

We very much hope that the fact of discovery of the lost "Catalog of Human Souls" will give humanity an opportunity to gain "Salvation"; "Salvation" for both individuals and the entire human race. Although on the other hand, it does not matter, from our point of view, through which social institutions the Catalog will be returned to civilization. More important is that each person now has an opportunity to find his real, natural, normal self, make himself in accordance with the "Divine Plan," and save himself like Noah and his family.

The legend of Noah, from our point of view, is quite noteworthy. Noteworthy in the sense that, for humanity to exist again as a biological type after the end of the Great Flood, and to exist within the "Divine Plan," the Creator ordered Noah to take "pairs of all living creatures" into the Ark. The legend of Noah's Ark, from our point of view, is one more proof that natural objects are etalon carriers for a human being, and natural images—genetic code as the life program of *Homo sapiens*.

Also, according to legend, Noah and his family, lived in the Ark for quite a while, and they were not just waiting for the end of the Deluge—they supported life of all the living creatures that were on board. However, it is possible to take care of natural objects only if you know them well. Otherwise, a canary will get fed some meat and a dog some millet; snakes will be kept in cold temperature and polar bears will be warmed by special heating devices, although temperature of -50°C is normal for a bear, but at a temperature of +15°C, it begins to overheat. It cannot be explained to animals, especially wild, that, figuratively speaking, pizza and chips with a Coca-Cola are useful and good, as it is impossible to persuade *Homo s.* to live in any way other than that, which is recorded in their subtype programs—they require their own, special conditions, and an animal either gets them or dies.

Therefore, in our opinion, the legend of Noah is a story about a human who went through the Way of Knowledge of natural images created by "God." Noah lived not some measly 70-90 years, but much longer, about 1,000 years. Therefore, the moral sense of the legend of Noah is that humanity has a chance for salvation from death and a long life only if people study themselves through natural images as etalons.

*****

We suppose that since it is not secular institutions, but religion that deals with an object called "soul"—religious institutions must provide answers to all questions relating to "Soul" as such, as well as to the "Soul" of every individual in particular. It is religious figures, and not psychologists, physicians, nutritionists, sexologists and other secular professionals who must provide people with recipes of life on all the six factors: intellectual, physical, nutritional, emotional, sexual, and environmental. We think that religion, not just secular science must be the source of knowledge about a human and the world; perhaps not all, but at least the main—who a human is and on what basis and for what purpose he functions. Religion, not scientific and technological spheres, must be able to solve all human problems since a human is an object created by "God," and therefore is a natural, not cultural or technogenic product.

Not to mention that showing humanity the true "Way to God," providing recipes of Spiritual Development to establish connection with the Creator is the task of religious, not secular institutions. This objective is inbuilt in the very concept of "religion" (from the Latin word 'religare' – to bind, to tie, to connect to something, to reunite).

As far as we know, the main mission of religion is to help each person restore the broken connection with "God," to reunite with Him, as it was in the Golden Age. Science can think all it wants that the story of the Golden Age, present in mythology of almost all nations of the world, telling about the blissful state of humanity living in harmony with nature and "God," is just fiction, utopia. Probably, science holds this opinion because it is unable to offer anything that could return humanity to this state. To science, at its current level of development, this is really a utopia. However, not to religion, which with the "Catalog of Human Souls" can return humanity to the Golden Age, make it a reality.

Logically, if the main objective of religious institutions is to connect a human with the Creator, to show humanity the "True Path," and since we have the "Catalog of Human Souls" as a very detailed reference of true life recipes for each person, we are able to help religious institutions realize these objectives in practice. Indeed, due to the discovery of this source of knowledge made by Andrey Davydov we have answers to questions of any individual: "Who am I?", "What is my purpose?", etc., as well as recipes on how to restore the lost connection with "God" for anyone who wants to do it. Regardless of a person's religious beliefs, race, nationality, gender, social status, etc. We can provide this information in the form of lectures, consultations, and so on to all religious institutions and their congregations.

We are also ready to contribute to the process of implementation by religious institutions their other objective, which to date has not been solved—struggle with sins. According to our research of the ancient treatise

Shan Hai Jing, a "sin" is only one thing—people's acceptance of artificially invented images as etalons instead of the images created by "God." And, we can help religious institutions since we have the source of information about natural images of any person. We are willing to provide information about natural images of any person, regardless of which religious confession he belongs to.

*****

Religious institutions and confessions can make any decision, and accept our assistance or not. However, as practice of the past millennia that led religion to a long-term crisis has shown, without the "Catalog of Human Souls" they are unable to achieve any of the above objectives, as humanity has grown from the time when it could live by faith alone.

To have faith modern people need arguments in the form of real facts. A modern person does not live by faith—he seeks explanations, answers to his questions, information. Today the term "blind faith" is an indicator of low level of development of individuals, and is no longer a common stance of modern people, even those who are not well educated. This is a peculiarity of the epoch in which we live. Not getting any explanations from religion on how a human and the world in which he lives are arranged, without evidence for the existence of "God," modern *Homo sapiens* concludes that religious institutions are nothing but a bunch of fraudsters who are only interested in money and power. For a modern person, even if he declares that he believes in something, faith as such and everything related to it ("belief", "trust", "gullibility", "superstition", etc.) are connected only with fraud, and that is quite reasonable.

Since the "Catalog of Human Souls" is the proof of the existence of "God," and is an explanation of how a human functions, it can become that missing link, which will help religion overcome negative attitudes towards religious institutions, and return "Faith." Especially since there is no need to believe in the Catalog, as any person can test the information from this source himself or herself in order to make sure. Note that in order to have faith it is necessary to make sure, to test, to ensure, and that is why, from our point of view, "True Faith" is Knowledge. And, when a person knows—he is sure, he trusts; in other words, he believes. Thus, "Faith" as a religious concept can be filled with real content; content that is understandable and accepted by a modern person. Because, from our point of view, offering modern people to believe in something, is like offering them to use a stone ax instead of modern technology. It is just as unconstructive, ineffective, as it is meaningless.

We feel that the tradition of "True Faith," which actually is Knowledge, should be revived. Everywhere. Globally.

If religious institutions will be able to do this, religion will take the leading place in the life of any human community, regardless of which race, nationality or religious confession that community belongs to. From our point of view, it is possible to do this if religious institutions return the "Catalog of Human Souls" to this civilization. Then, religion will finally really help humanity see the "Light of God," know the "Divine Truth," and go through the "Divine Path." It is hopeful that through the acquisition of this "Light," the Light of Knowledge, humanity will receive "Salvation," avoid another "Great Flood." From our point of view, in order for this to occur, all that needs to be done is for each person, whether a believer or not, to know that the "Catalog of Human Souls" exists, and to learn about his natural potential from this Catalog. Then, humanity will not need to reproduce such a huge number of artificial images since everyone will know that the richest source of information about their own, natural images exists, and that every interested person has access to this source. People will become as "God" created them, mutations will stop, and the Creator will not need to destroy this civilization. Salvation of humanity, from our point of view, can occur only if the will and the creative potential of humanity will be directed in the right way, in the direction towards conformity with the "Creator's Plan." And, this is the only way of salvation.

People must learn about the "Catalog of Human Souls," that "God" exists, that "Soul" exists. They must learn what each one of them has his own, personal "Soul," and that this "Soul" is knowable. If instead of believing, they will know for sure that they were created by "God," that they are the children of the Creator, then they will no longer doubt that they should humbly follow the "Divine Will" of their "Father." Then, the world will finally become the way "God" wanted to see it. Is that not the main objective of religion and church?

*****

We also think that if religious institutions return the "Catalog of Human Souls" that was left by our Creator to humanity, it will save them as existing (and demanded) social institutions. And, religion will finally find a proper place in human hearts, minds, in human life.

Despite this, we do not insist on participation of religion in this process since our Creator has already taken care of everything. As already mentioned above, in the event that religion decides to continue to function based on principles on which it functions now, meaning without use of the "Catalog of Human Souls"—our Creator has given each person an ability to communicate with him directly, through the use of natural images of one's program, "Soul."

PART 13

RESPONSIBILITY OF RELIGIOUS INSTITUTIONS TO THE CIVILIZATION

It should be noted that humanity is left with a choice: to incarnate the "Will of the Father" by following "His Plan" or not. Forcing humanity to make the right choice, apparently, was not part of the "Creator's Plan" because this would be contrary to His decision to give free will to a Human. Probably, it is time for people to take care of themselves. To do this, from our point of view, it is necessary to decide whether to be with "God" or to continue to live as mutants. Consequences of both decisions are already clear.

From our point of view, the sacred responsibility of religion is to help people choose life, Salvation. Since, as shown by our study, humanity, unfortunately, is no longer able to make the correct choice without help. It has degraded too much during experiments with artificial images. Mass craziness and other facts indicating degradation of the existing species *Homo s.* are already evident on all continents. Using the language of religion, by their sins humanity had brought itself to such a level of incidence that it is no longer able to hear not only callings of religious institutions or "Voice of God," but also the voice of their own mind.

According to our research, to date humanity has brought itself to the state of an insane animal that is unable to stop by itself, to end the processes of destruction and self-destruction, to cease to violate the "Divine Laws." Even a clear lesson with Jesus Christ that was given to them, humanity is unable to interpret correctly, to draw appropriate conclusions for already more than 2,000 years. Therefore, from our point of view, it is already meaningless to expect prudence from humans.

The future of humanity—to live or to die—largely depends on representatives of religious institutions. Although it has been and will be like this at all times because the essence and foundation of religion is to guide people to Light, to give people the "Divine Wisdom," and thus to help save the life of civilization. However, religion has every right to not do this. Although in this case, we, as researchers of the book left to people by the Creator, who are able to give information about how this civilization can survive, will feel a bit sorry for civilization. However, we hope for prudence of representatives of religious institutions, as who knows if not they that when the time comes, the Creator will spare no one, and in this case not only shame awaits them, but also destruction together with congregation.

It is likely that like last time, the only left alive will be those who honor the "Will of God" and follow it. Such people already exist today. It is likely that this time one of them will be tasked with, like Noah, saving the DNA collection for next generation of humans.

*****

However, there is another danger in addition to destruction of the existing population of *Homo sapiens* through utilization mechanisms of nature. And, we are obligated to inform about it. The "Catalog of Human Souls" is already in active use for about 20 years to date. If religious institutions will not wish to participate in its widespread and proper use, it is possible that at some point the Catalog will again be used by a limited group of subjects whose objective is not Spiritual Development, "Enlightenment," and "Salvation" of humanity, but only acquisition of unlimited power and total impunity.

Since, as it was mentioned above, influence of people that the use of the Catalog of human population provides is always very effective, but absolutely unnoticed by the subject towards whom it is directed. And, it is impossible to trace or even to notice this influence by a person who is not familiar with the Catalog. Such possibilities of influence always attracted and attract attention of social groups related to power structures, politics, etc. Also, this ability is regarded as very attractive by each type and level of criminal elements, as the use of the Catalog with the wrong goals cannot be tracked and stopped by law enforcement authorities of any country in the world. People become absolutely defenseless against evil will of those who not just might want to, but are already actively trying to use the knowledge of mechanisms of human psychophysiology for evil purposes.

Given that the "Catalog of Human Souls," like a knife, can be used to people's benefit, as well as to harm people, we consider the question of who will get their hands on this source of information, this tool—extremely important.

*****

Unfortunately, we already have experience with use of information from the "Catalog of Human Souls" not in the way, in which this source of information must be used. In the recent past (2000-2010), the "Catalog of Human Souls" was practically expropriated by a group of employees of the Federal Security Service of the Russian Federation (FSB, formerly KGB). Upon learning of the existence of the "Catalog of Human Souls" from the media, this group, together with figures at the highest echelons of power of the Russian Federation and security forces, became very interested in the possibility of obtaining from it any kind of information about any person of their interest. And, in gaining unlimited influence by using the tools described in the Catalog.

The aforementioned group of employees of FSB of Russian Federation requested information from the "Catalog of Human Souls" under the guise of ordinary, simple people for a long time. The fact that they used this information for, to put it mildly, nefarious affairs became known only later on.

In parallel, the special services of the Russian Federation by all means hindered informing of wide audiences about the discovery of the Catalog of human population made by Andrey Davydov; both in Russia, and in the international zone. Also, for nearly fifteen years, this group of employees of FSB of Russia has spread information through the Internet that was frankly false and defamatory of our research product and us. For this reason, the Internet has accumulated a mass of absurd fabrications, false information about the "Catalog of Human Souls," unflattering "reviews" posted by FSB officers disguised as ordinary people. In addition, during the time when we lived in our home country, the former KGB blocked all contacts with us, controlling all of our communications: home and mobile phones, computers, mail, etc.

Throughout this period, we have been essentially cut off from ability to communicate with people who were not related to the FSB and politics, including with colleagues in scientific spheres in different countries.

It is for this reason that today few people know about our product and us, and information about the Catalog of human population that is available on the Internet, is mostly false and slanderous. It is for this reason, and for no other that people who do not know us personally, have no way of contacting us at this time. Since a group of ex-KGB employees is still hunting the "Catalog of Human Souls," and us—its researchers, we still do not have the opportunity to put on public display on the Internet information on where to find us and how to contact us.

Andrey Davydov, the author of the discovery of the "Catalog of Human Souls," has always been strongly opposed to use of this source of information for the purpose of evil and lawlessness. We, the researchers of the Catalog of human population, always held the position that each one of the 7 billion people living on earth has every right to have access to information from this source of knowledge. Also, we have always tried to ensure that the Catalog of the human population gets used exclusively for the benefit of people, for their development, and primarily for their Spiritual Self-Development. It is for this reason that the aforementioned group of employees of special services of Russian Federation has tried to kill us using small doses of toxic substances for several years. For this reason, in 2010 we were forced to leave our homeland and were granted political asylum in the United States of America.

Despite this, the interest of Russian special services in the "Catalog of Human Souls" has not cooled. Therefore, we would like twice as much for the masses to learn about the "Catalog of Human Souls" as soon as possible. And, not just to learn, but also to start using the "Catalog of Human Souls," making their lives more correct, better, brighter, more pleasant, productive, and effective. That is the reason we want this book to finally turn up in good hands.

*****

We are well aware that any social institutions of any country in the world can try to follow the same path as the FSB of Russia, and assume that since the Catalog has been found, any specialist in the study of ancient texts is able to decrypt it. However, we also know that this is not possible because more than fourteen years of experience has shown, that despite the fact that the Russian special services have enough analysts, very high level specialists in various fields of scientific knowledge—none of them were able to decrypt the texts from Shan Hai Jing.

Meanwhile, the Catalog is not yet fully decrypted. To date, only one third of the Shan Hai Jing has been decrypted. And, only of the first program level described in the Catalog of Mountains. Also, as experts in the field of non-traditional psychoanalysis, we know very well that people's lives directly depend on the correctness of decryption of text from this ancient book. Since any inaccuracies, errors in decryption of this text negatively affect the state of psychophysiology of *Homo sapiens*. Our practice of about twenty years showed that among the possible consequences are severe psychical disorders, usually without an option for recovery, as well as very serious somatic disorders. Illiterate decryption of texts of Shan Hai Jing gives purely negative results instead of giving positive. Therefore, we, as the only specialists in the world in the study of this ancient source of knowledge about human psyche, are entirely responsible to ensure that humanity receives a qualitative research product; a product that, as we think, will be in use as long as this civilization continues to exist.

PART 14

"MEMENTO MORI": "SOUL" OF A HUMAN IS NOT SECOND HAND

> "Respice post te! Hominem te memento!"
> ("Look behind you! Remember that you are a human!")
> Tertullian (Apologeticus, Ch. 33)

It is a shame to note the obvious fact that in the XXI century people do not understand the most basic things. The level of people's knowledge about the basic questions, such as "life" and "death" is so low that in this sense, a modern human is not at all different from a Neanderthal. Facts to support this statement are obvious. Both, figuratively speaking, raise their eyes to the sky, hoping to see the "souls" of the deceased there. It is no secret that even today, on all continents, there are fairly widespread beliefs in immortality of the "soul," in possibility of bodiless existence of a human, in reincarnation, in resurrection of the dead.

From our point of view, in order for religion to find its dignified place in the minds of people, it makes sense for religion to stop playing the game when it claims to be the source of universal wisdom, and even makes some requirements of a human, but then hides behind the door: "Religion is just a pass-down, folklore—believe it or not, as you want." From our perspective, it would be more appropriate for religious leaders to take part in providing people with reliable information about some of the questions that are directly related to religion.

*****

For example, one of the functions of religion is to explain to people subjects related to death, and what happens to human "Soul" after this event. We found some interesting details on these topics, and are willing to share some of them. Since lack of information is the worst for a person because it always leads him to mistakes, and often irreparable. And, as it turned out, death of people of the level of those living in this civilization is the final "point" in the book of their life.

*****

"But as it is written, Eye hath not seen, nor ear heard, neither have entered into the heart of man, the things which God hath prepared for them that love him." (1 Cor. 2:9)

In regard to the so-called "Eternal Life," for example, Christianity says that for those who have faith, love "God"—this life will be so blessed that it is impossible to imagine or describe it. However, it does not say what it will be exactly (which is typical of religious sources, famous for lack of specifics). However, some of the ancient books have revealed different information about "Eternal Life." There, it states that immortality awaits only those who really, fully embodied "God's Plan," reached the level of "Image and Likeness," and made themselves Humans. Everyone else can only dream of "Eternal Life" as immortality (resurrection) of anything, the "soul," or the body. However, these dreams will not bring any results.

As it turned out, immortality, possibility of resurrection, is an ordinary, normal state of a Human as a creature born as a hard-coded and rather primitive (compared to the inborn potential) bio-machine, but that passed and mastered all seven levels of programs of development, described in Shan Hai Jing, and made a Human of himself. By the way, Jesus Christ has demonstrated this by his "gymnastic exercises" on the cross—He has risen, and showed that for him personally death does not exist. Jesus Christ showed people that if they will become the same as Him, the "Son of God," then they will have exactly the same abilities: will replicate bread, walk on water, heal and resurrect the dead, and shall live forever. (Only, humanity somehow misunderstood Jesus Christ, namely, decided that the "Son of God" will do it all for them.)

Incidentally, according to one of our untested hypotheses, a huge percentage of human DNA, which at one point was determined by geneticists as "genetic junk" may actually be a record of all levels of development of *Homo sapiens*, of which there are seven, individual for each subtype, and that is a huge amount of information. However, this is just one of our suppositions, and must be verified in line with interdisciplinary research.

To an individual who has attained the status of a "Human," such natural mechanisms as "life" and "death" are just tools, like a hammer or a drill. He is not impacted by these tools, but uses them freely. This is the difference between abilities of a Human and abilities of people of this civilization, who can only dream of being immortal, but die in reality.

According to ancient sources, Human's ability to live forever is as natural as the ability to walk, to speak, to read, and to write. The way to acquire the ability to be immortal, according to our data is the same as with speech, writing, reading, walking, etc.—to learn. To learn to live according to your program, which upon mastering is replaced by the following program, then the following, and so on. This is the process meant by the words "went through all seven levels of programs of development" and "made a Human of himself." This means that not in his personal dreams, but in reality a person acquired complete similarity with our Creator, became "the Image and Likeness of God," and, most importantly, that he is able to demonstrate his divine, extraordinary abilities at any time. (From our point of view, this is different from when people claim possession of some extraordinary abilities, according to the principle: "no one will check," but he or she is unable to demonstrate them.)

*****

According to our data, immortality is an ability only of a being with a status of a "Human." Representatives of other levels of development of human beings are destined entirely different fate by the Creator. It does not make sense for them to even dream of immortality of not only the body, but also of their "Souls" since, according to information from Shan Hai Jing and other ancient sources, immortality is not a bodiless existence of "soul," but a full physical human existence. From our point of view, this specification is also quite significant in light of existing beliefs in civilization.

It is stated in the ancient books that those people who failed to achieve the level of development of a "Human," should not expect "Hell" or "Paradise"—they can expect only a garbage dump called a "graveyard" or a "crematorium," where human bodies are stored like old, unused equipment.

"Souls" of these dead do not go to neither "Purgatory," nor "Paradise" and not to "Infernal Boilers of Hell." As no longer used, old versions of operating systems, like Windows 95, they go, so to speak, into "world archive," where they will "gather dust on the shelves" among "Souls" of other people who did not follow the "Plan of the Creator" during their lives. This is so because from the standpoint of nature and the Creator, "Souls" of this quality are worthless, useless—they did not grow, did not develop, did not learn, did not transform, and therefore, there is nothing interesting recorded in these "files."

However, a second chance will not be given to these "Souls." According to information received from the ancient sources, the Creator gives life to any representative of the biological type *Homo sapiens* only once. In this sense, all living natural objects are equal: people, animals, plants, etc.

Human is given life only once; life as a chance to make himself a Human—a being equal to his Creator based on real, observable indicators and abilities. However, if a person did not use his chance and spent his life living in his fantasies, where he was equal to "God," spent it on pleasure and entertainment, then within a certain period this individual gets utilized through natural mechanisms.

It probably makes sense to explain in short what we mean by the words "pleasure" and "entertainment" because we do not mean that what is usually meant by these words. We do not mean that, figuratively speaking, a person who eats only pastries or sits on a deck chair by the pool with a glass of Martini all day every day, etc. We partially demonstrated the meaning behind these terms from the standpoint of the ancient sources in the section about living people with dead "souls." People who live on artificial images, enjoy not only pastries, but also, for example, the process of slow daily suicide, or the process of killing another person with their own problems and troubles, with destruction and chaos, with confusion in their own affairs because they do not understand anything, and so on; and, in general, through that perverted life model by which they live: "was born—studied—got married—worked and brought up children, then grandchildren—died." Despite the fact that this model is perceived by people of this civilization as normal, from the standpoint of ancient knowledge about potential and mission of a human, this is not even animal life, but a life of a human-mutant because such life is normal for an animal, not for a human. Life of a human-mutant, from the point of view of these sources, is a life where there is no studying of his subtype program from the "Catalog of Human Souls"; no production of qualitative creative products that were created not on the basis of fantasies of a madman, but based on natural images; no learning; no goals that each person must strive for and reach; etc. As a result, such life is, in fact, outright blasphemy because it was said that He created us in His image and likeness.

People entertain themselves by business and other social processes, or by destroying their lives, their own selves, and all that is around them. However, they are firmly convinced that, as one Russian saying goes, if a person planted a tree, built a house and raised a child, then his life did not go in vain. This is not true. Therefore, we call these "a tree—a house—a child" by different words: "pleasure" and "entertainment." In other cultures, the same values are recorded by other images, but that does not change the meaning because apart from pleasure and entertainment people do not get anything for their own psychophysiology, just spend their only life. Moreover, they waste it since this kind of life is not life, as created for a human by the Creator by way of his subtype program. However, more importantly, this way of life does not make a human—a Human.

With realization by a human of a life script "a tree—a house—a child," his body, and his personal "soul" fall under the natural mechanisms of utilization; because from the standpoint of the Creator, he is a rotten (or dried) seed, and from the standpoint of nature—a worthless object that does not fit into the overall natural system.

*****

"Soul," as a natural program of a subtype, exists regardless of life or death of separate individuals. It is quite possible that this is where the belief "Soul is immortal" came from, and another Russian saying fits very well here: "He heard a ringing, but doesn't know where it is." Yes, indeed, "Soul" is immortal, but human body is not a thrift store, where you can put any junk as an "operating system," and it will work. Human "Soul" does not fall under the category of second hand—"wear it for a while, and give it to someone else." Nature does not need junk. Nature of our planet is a factory with incredible powers, which operates smoothly, and without interruption "gives products," including natural "operating systems" for any natural objects.

To put it in computer terms, "souls" that were "previously used" do not get "installed" in a body of a newborn. The Creator created nature in such a way that every day it is able to create a fresh "operating system" for those born on this day. Although when considered as models, as matrices—these "souls" exist as long as life exist on this earth.

Therefore, it is meaningless for people to hope that the "soul" of a deceased person will go into a new body, as that "soul" will not "reincarnate" even in a worm or a bush because every natural form is born, using the language of modern technology, with "factory settings," with a pre-install, with "presets" (Eng. 'preset' – presetting, a set of configuration parameters of electronic equipment or software specified at a factory at the time of manufacturing). Even more so, one should not expect that the "soul" of one person will "dwell" into the body of another, as from the standpoint of work of natural mechanisms this is impossible. An "operating system" (in religious terms, "soul") for newborns is always "fresh," only "straight from the oven." As it turned out, everything else are simply primitive human fantasies, born in the absence of knowledge of natural mechanisms.

*****

On planet Earth live, so to speak, a large number of clones, duplicates of natural programs: birds, animals, fish, stones, metals, plants, humans, etc. However, everyone, animals and humans, is also given that what is commonly called by the word "individuality."

On the one hand, individuality (from Lat. 'individuum' – indivisible, individual) can be looked at from the point of view of an existing definition in civilization, meaning—as a set of properties and characteristics that distinguish one person from another; as originality, uniqueness of psyche and personality. However, on the other hand, from our point of view, this definition is vague. From the standpoint of knowledge from the "Catalog of Human Souls" and other ancient books connected to this book, individuality is the sum of a person's body with the common "soul" of the whole subtype, plus personal life experience of the individual.

As it was already mentioned above, subtype program ("Soul") is one for the entire subtype. It is like this for humans, and all other natural objects. However, in combination with a particular body, with the environment, which includes living conditions, upbringing, program characteristics of parents and teachers, education or lack thereof, etc., with individual experience of body and "soul," obtained in the process called "life"—this is individuality. Although, as already mentioned, no body that was given by nature, no upbringing, no education can change the natural mechanism called "program" ("soul"). The environment can require some program qualities of a person, but not others, or might try to block, deform or even break them.

However, the subtype program will not change. For example, a mineral can be broken up, but it does not cease to be a mineral; a plant or an animal can be destroyed by wrong conditions when held in captivity, but it does not cease to be the carrier of properties of its subtype.

Every living creature has individuality, and pet owners know this very well. Having gotten an animal of exactly the same breed, for example, instead of the one that passed away, they notice with surprise that it is not an exact copy. This is individuality.

However, individuality does not change the natural program of a subtype. Therefore, two animals, despite the fact that, relatively speaking, each one has spots on the skin located in different places, they will live and function in the same way. Stones also exist in completely different shapes, sizes, colors, may be located on different territories, but they are still stones, they do not turn into, for example, metals. Therefore, regardless of how this or that representative looks, what environment he grew up in, what personal experience he has—all members of one human subtype are the same, "like ball cartridges."

It is very likely that people confused the concept of "soul" as a subtype program and "individuality." And, they decided that their individuality is just as immortal as subtype's "soul."

With the loss of body as an integral part of individuality, without which individuality ceases to exist, the "soul" as a natural matrix of a subtype continues to exist, but the individuality of a person gets utilized entirely; entirely as it existed prior to his death: the body of a person, his subtype program and his personal experience.

*****

Unfortunately, we have not yet found information about what happens to "Souls" of those, who followed their program "Path" to realization of the "Creator's Plan" in respect of their own selves, but did not have time to achieve the status of a "Human." Therefore, we cannot report anything on this subject yet. A person who is familiar with his natural program from the "Catalog of Human Soul" is very different from those who are not familiar with their natural programs. He differs at least in that he clearly knows who he is and what the purpose of his life is. He does not need to ask society: "Who am I? What am I? What should I do?" He does not need anyone's judgments, advices on how to live, or anyone's opinion about his natural qualities, his actions, his way of life. He has completely different algorithms of life, worldview, interests, aspirations and objectives, a different attitude towards nature, society, people, and himself. He knows whom to be, what to do and what to strive for, and, notably, this knowledge is not born in the space of his sick fantasies, but is taken from the real source that anyone can see, feel, read. From out point of view, life of a person who at least tried to realize the "Creator's Plan" in respect of his own self, is more worthy of a human being than life of an animal. At least while he is alive, this person lives as a human, and not based on the principle: "eat—go to the bathroom—reproduce." However, this is our personal opinion, and we do not insist on it. We hold the position that each person has the right to live like an animal, but each person also has the right to not live like an animal.

*****

This information that was received from the ancient sources, from our point of view, is very important. Knowing this, people will not have illusions that they will be able to live many times as in a computer game; that they will live forever, moving their "soul" from one body to another; or exist forever in bodiless state "somewhere out there", "in the vastness of space." From our point of view, it makes sense for people to know that this is not so. The myth that you will die, but your supposedly immortal "soul" will live on is extremely harmful because, in this case, people do not care about their body and "soul" during their lives as it is required, and, as a result, end up, figuratively speaking, in a garbage dump. Does it make sense for a person to believe in a beautiful fairytales if the price is his life, which will end forever?

As our practical research has shown, people who are not familiar with their natural programs do not know even close how to take care of body and "soul," how to handle them, even though these are specific values with hard-coded parameters. Without knowledge of these parameters, it is impossible to provide required conditions for their work and safety. Human body and "soul" require exactly the same literate and careful handling as a piece of newest, super complicated, very expensive technical equipment.

Moreover, work of human psychophysiology is by far much more complex than any such device. A human, as a mechanism, works in an even more strict way: if at least somewhere, somehow, at least something disrupts the required conditions of content or use (no matter due to ignorance or for "entertainment"), then his entire psychophysiology gets deformed and begins to fail very quickly.

Any technical device comes with an instruction manual, implying that it is impossible to intuitively understand how it works, and even more so in relation to a human. To date, *Homo sapiens* living in civilization without knowledge about their own device from the "Catalog of Human Souls" remind of Neanderthals, who got their hands on a modern super-powerful computer. And, all they do is try to open coconuts with this computer, or clobber all the surfaces around with it, while observing how long it will last. The results of such Neanderthalism, and, Neanderthalism without quotes are always impressive and very sad. In fact, they are analogous to trying to use a washing machine as a microwave by putting sandwiches in it, cleaning a refrigerator with a hot iron, ironing clothes with car tires, or washing socks in a coffee grinder. On the other hand, everyone knows that a coffee grinder is much easier to fix than human body and psyche. Everybody knows from personal experience that "breaking is not building," and any violation of a human body, psychical and somatic, is treated with in a long, difficult, unpleasant, painful way, and, more importantly, not always successfully. And, everyone knows that at a certain stage of "breakdowns"—doctors and medications can no longer help a person.

Possibly, news that reincarnation, including of the "soul," does not exist in nature is not from the category of "pleasantly-interesting" that people of this civilization are used to, like trained doggies. However, we hold the position that people should know this, should know that this is so and not otherwise because, then they at least will have an opportunity to plan and spend their lives in some other way. Otherwise, engaged in pleasant self-deception, in fact, people lose their only chance in the form of life. And this, from our point of view, is not just unproductive in terms of use of human resource, but is also terrible. We also think that hiding this information from people is indecent and criminalistic. Everyone has the right to know that as long as he is alive, he must take care of his own self, know himself in order not to break himself. Unless, of course, one wants to end up "in the dump," where there is only the eternal nonexistence and eternal oblivion.

*****

In addition to fairytales about immortal "soul," humanity loves tales that at some point someone will appear, who will save it and solve all the problems that it created "for fun" for itself. World culture confirms it. The idea of Messiah exists not only in Judaism and Christianity that came out of it, but also in Buddhism, where the savior is considered to be the future Buddha Maitreya, and in Islam, where there is the idea of Mahdi, and in Hindu (Kalki Avatar, etc.). However, according to the results of our research, it is useless to even hope for this, let alone rely upon it.

"Second Coming" will not occur because all has already been demonstrated to humanity. For instance, the biblical example: "Son of God," Jesus Christ descended to earth, and by his personal example showed people what they should be, and what they can become, if they follow the "Plan of the Creator" in relation to themselves. From out point of our view, Jesus Christ does not remind of a clown who will repeatedly entertain humanity with similar "shows." Now either humanity will remember the lessons of Christ, or it will once again be washed away from the face of the earth, and the Creator will inhabit this planet with the next generation of humans. Who, very likely will be more understanding and humble, and follow the Will of the Father—for every person to really have the same qualities as His Son, Jesus Christ. "Jesus Christ is the center of God's plan." (Genesis 1:1, Acts 17:24-29, Psalm 103, Isaiah 45:18; Excerpt was taken from Russian versions.)

We think it makes sense for people to know about this too. Since, from our point of view, this gives every person who does not want to end up "in the dump" an opportunity to start doing something in the direction of his own salvation. Possibly, this behavior would be much more productive than passive waiting that one day Messiah will come, and wars, illness, and death will stop, and all human problems will disappear, and there will be universal peace and prosperity, and people will get an opportunity to enjoy peace and harmony, and will finally be able to devote themselves to knowing "God" and to spiritual self-development. It is a beautiful fairytale, no disagreement there, but the result of believing in this fairytale is not so beautiful: wars, diseases, murders, deaths, violence, suffering, and other disasters of humanity continue.

According to information that we have, humanity must earn all the benefits it expects to obtain through the Messiah by people's own efforts, their own work, and no one is not going to do this for *Homo sapiens* because this is what Spiritual Development is in practice, without games, without speculations. The Creator gave huge potential to each person. He has implanted in every human being the ability to organize any environment for themselves. The proof for this is the "greatest achievement" of this civilization—creation of "Hell" in "Paradise." From our point of view, there is every reason to think that if humanity by their incredible insistence on stupidity, laziness, idleness, fruitless dreams was able to organize themselves "Hell in Paradise"—it is capable of solving problems like wars, curing people from illnesses, and having a "blooming garden around them."

Each person will decide for himself whether to say every day, like a Jew, "I unconditionally believe in the coming of Messiah, and although he is delayed, I still will wait for him every day," while continuing to suffer every day in his personal "Hell," or start doing that what he expects from the Messiah with his own hands. However, people should know that today there are those who are already studying and mastering the "Catalog of Human Souls," and, figuratively speaking, already "moved" from personal "Hell" to another, more pleasant place for living. And, currently this is not much more laborious than putting out a glass full of wine, an empty plate and silverware, and leaving the door open in anticipation of arrival of prophet Elijah, harbinger of the coming of the Messiah. Based on results, this is much more effective. However, it is likely that some are quite fine with doing nothing and having nothing—"to each his own."

*****

From our point of view, in conclusion of this part, it would probably be reasonable to also break the myth that the comprehension of the "Divine Plan" is an option that exists for humanity at its current level of development. Since, to our knowledge, this is not so.

It would be very good for humanity to get rid of the illusion that it is already very well familiar with the "Divine Plan," and begin to explore sources of information on this subject in order to have at least some idea about it. The Creator intentionally left these sources of information to humans. Probably, this prompted the Creator's decision to leave Shan Hai Jing to humanity since humanity, at the initial stage fails to comprehend this on its own. Being at a level of a helpless baby, all that humanity is capable of is dreaming up something that is called philosophical and religious wisdom in civilization, but has absolutely nothing to do with the Creator and "His Plan" for humanity. From the standpoint of information from the ancient sources, and, above all, from Shan Hai Jing, such "wisdom" does not hold water. And, our personal opinion has nothing to do with it, as these are objective facts.

From our point of view, it makes sense for humanity to stop thinking that it understands the "Creator's Plan," at least because it will only make an already difficult situation worse. Moreover, using the language of religion, this will angry "Our Father." As dreams of people who live on the basis of artificial "operating system" work only in one direction: creation of artificial images. And, this is very dangerous.

Being at a state of unwise, helpless and frankly unhealthy child, humanity is simply not able to understand the phenomena of such a level as the "Will of the Creator", "His Plan," not even if it strives to and very much wishes to. People seek "God," this is natural and inherent in every person. However, a small child is unable to understand the origins of behavior of adults, and humanity, at its level of development, is unable to understand the Creator.

From our point of view, religious institutions could explain all of this to people. Our Creator did everything for our own good, but he created people in such a way that they are obligated to do something themselves as well for their own well-being.

And, this is not our opinion. This is the result of our research in the form of decryption of the ancient Book that in civilization turned out to be the only source of information that was not created by people.

EPILOGUE

"Spirits edified, demons instilled…"

"Behold, I stand at the door and knock. If anyone hears My voice and opens the door, I will come in to him, and will sup with him, and he with Me." (Rev. 3:20)

*****

We are hopeful that the brief information that fit into the format of this book about structure of *Homo sapiens*, human "soul" and the "Catalog of Human Souls," will help people to at least get the answer to the question about the meaning of their lives. And, they will not have to, like the author of Ecclesiastes, ask themselves strange questions like: "What is the benefit to a human from all his work, what does he work for under the sun?" Since the answer has long ago been recorded in the ancient sources that we discussed, and our readers already know that they state that every *Homo sapiens*, first of all, must work hard to make himself a Human.

We also hope that the information about where artificial images come from, and the implications of following "programs" implanted in them will help people learn how to distinguish useful from harmful. Since if one does not, then images from books, movies, songs, paintings, and the Internet, which soak in as food for the mind, begin to serve, using the computer language, as program applications according to which a person begins to live. We think it made sense to write this book for this alone.

We think people will begin to understand that while reading, for example: "Vanity of vanities, saith Ecclesiastes, vanity of vanities; all is vanity" (Ecclesiastes 1:2), knowing nothing about the mechanisms of artificial images and their impact on human psychophysiology, a person starts to live and act by this program. And, in the end all of his life, all his actions, efforts are in vain, useless, do not bring results. Since synonyms of the word 'vanity' are concepts such as "emptiness", "anxiety", "paltriness", "commotion", "haste", "bustle", "empty chores", "scurry", "trumpery." (By the way, does this not resemble the life of a modern representative of this civilization?) In the end, a person himself turns into a useless subject, a nothing sooner or later, as everything that he does not have any meaning or benefit. In addition, 'vanity' is also perishability because among synonyms of this word are words 'ashes' and 'dust'. And, 'perishability' means mortality. Therefore, a person who naively "swallowed" this statement thereby turns on the program of self-destruction.

This is how artificial images work.

Of course, everyone has the right to consider the author of Ecclesiastes (Ecclesiastes is the name of a book of the Old Testament that in the Christian Bible is placed among books of Solomon) an incredibly wise man. No matter whether the author was King Solomon or someone else. More important is that, in terms of information from books that are dozens, even millions of years older than this work, he is simply a naïve, confused subject who knows nothing and understands nothing. Since only a confused person can draw a conclusion like: "I have seen all the works that are done under the sun; and, behold, all is vanity and vexation of spirit" (Ecclesiastes 1:14). Only a person who knows nothing about his own self and the world might think he knows everything.

Or, for example, another well-known statement by the same author: "Who increaseth knowledge increaseth sorrow" (Ecclesiastes 1:18). From the perspective of knowledge that the main function of *Homo sapiens*, according to the "Creator's Plan" is to learn—it is difficult to utter a bigger preposterousness. People who trustingly "eat" these sayings can be sincerely pitied. If everyone shared this position, today science would not exist, and humanity would continue to run around with a stone ax. However, more important is that the acceptance of this position makes a person into a madman because information is food for human intellect. Blocking his access to perception of new information, one lives only in his imaginations and memories, and, as it is called, "goes crazy" because his intellect stops receiving nutrition, exactly the same as if he would have stopped feeding his stomach. However, on the other hand, multiplication of one's knowledge by consumption of fruits from the dreamland of another philosopher-dreamer is not only pointless, but also harmful to health. After all, a reader, whether he likes it or not, whether he agrees with it or not, begins to live according to pictures "painted" by the author because a human is a living machine that works on programs, both natural and artificial.

Since all images presented in literary and other works of *Homo sapiens*, such as painting, music, cinema, are artificial images—life of naïve, unsuspecting consumers of these products begins to "go downhill" very quickly. There is a very interesting effect: everything that another "king solomon" "paints" for a reader (listener, viewer, etc.) at some point begins to appear in his or her life. Or, as they say: "comes true." Moreover, a person perceives this as confirmation that another author of "great wisdom" was right, when in fact, he as the consumer of artificial images has successfully implemented the program of actions that was recorded by these images. He did this with his own hands, and it is unimportant that he had no idea what he was doing and why.

The fact that people now have at least some idea of what an 'image' is and how it works, there is hope that they will be more critical of information that comes to them from the outside. Since they will know that it will all appear inside them like a sandwich, and their cells will start to consume this information, changing them and their lives not at all for the better.

We were happy to share this information. Therefore, it the beginning we stated that, despite the fact that this text was created as a new ideology of religions, we are absolutely indifferent to how representatives of religious faiths, institutions will react to this information and whether or not they will react at all. We wrote this book for people because we know that in each one of them a part of our Creator is implanted, a part of His potential. And, we want people to be able to realize this potential.

By the way, according to ancient sources this is what doing good means.

*****

As far as our participation in any kind of civilization processes—we are not interested in them. We are no longer interested in all that does not relate to us personally. However, by words "us personally" we do not mean that what is commonly meant by this; we mean that what does not apply to our psychophysiology: psyche ("soul," subtype program) and body since the only thing that applies to them is that information, those activities which are prescribed to each of us in the "Catalog of Human Souls." This is what is important to us. This is what we are interested in, and not something else.

Therefore, we ask all those who want to talk with us after reading this book, to not offer to us participation in non-constructive discussions, or, even more so, demand evidence from us. We are not engaged in providing evidence, we are engaged in our research studies, and evidence in this process appears on its own. From our point of view, nearly 280 published books with descriptions of people that anyone can test by comparing these descriptions with specific, real people, carriers of these programs is sufficient evidence that the Catalog of human population exists, and that we know how to decrypt this book. As for discussions, we are ready only for those that relate to the essence of the question, but in any case are not a type of "entertainment" when a person is just bored, or just wants to talk with a goal of feeling interesting and useful to someone, or to play a game "You are all fools, only I'm smart, and now I will prove it," to us this is all the same.

Also, we would like to ask to not make recommendations to us to study religious sources, supposedly referring to that we are not familiar with all of them, and so cannot offer a new ideology of religions. We are well aware of this trick of religious leaders and not only them. In our understanding, this belongs to the category of phenomena that we described in Part 10 ("Dead Souls": Human Mutants). As practice shows, it is completely unnecessary in order to understand the essence of processes occurring in the world of religion. Just like it is unnecessary to step on feces in order to make sure that it is not a chocolate cake. For us, those books that we study are quite enough to understand all the processes, phenomena that take place in society. To us it is enough to be experts in our field, and we do not claim anything else. We do not want to seem arrogant, we are open for any life processes, but not for dead processes, meaninglessness, which civilization proposes to engage in. We are simply familiar with arithmetic, and calculated that if we consider human life span as 90 years—that is even less than 800,000 hours. We value our time, and we have things to do.

*****

In conclusion we would like to say that we understand very well where the "legs grow out of" in any kind of mystery (see Part 10—"Dead Souls": Human Mutants), and are working hard so that there are fewer mysteries. However, we differentiate between mysteries as something that exists in nature, but is not yet known, and "mysteries" invented by people in order to hide the fact that they lack self-knowledge, lack interesting information and just want to attract attention. And, we think that people need mysteries from the first category. Therefore, we would like the words that stand under the Epilogue—"Spirits edified, demons instilled..."—to remain a mystery to our readers, a mystery that perhaps they will want to unriddle sometime, as well as other mysteries, an abundance of which were left to all of us by the Creator.

CONNECT WITH US

We said everything we wanted to say in this book. If someone wants to communicate with us personally, currently, the only way to contact us is via one of the Internet resources, where we sell information from the Catalog of human population: www.amazon.com. It is easy to do. Simply go to http://www.amazon.com/gp/help/contact/contact.html?ie=UTF8&isCBA=&marketplaceID=ATVPDKIKX0DER&sellerID=A1H58IR0HP7GHK; enter your login and password; select "An item for sale"; select the option "Other Question" (next to "Select a Subject"); click "Write Message"; write your message; and then click on "Send email" to send it. The second option to reach us is to contact the Human Population Academy (see http://www.humanpopulationacademy.org/breakthrough-discovery/contacts/) and your message will be forwarded to us. However, please note that if information that you want to communicate to us is very personal or confidential, it is possible to contact us directly only via Amazon, otherwise our response will be sent through the Academy.

ABOUT US

Special Scientific Info-Analytical Laboratory—Catalog of Human Souls was founded by Andrey Davydov. The laboratory is engaged in research and decryption of the ancient Chinese monument Shan Hai Jing, as well as other ancient texts, and creation of the *Catalog of human population*. The technology of uncovering individual structures of psyche of *Homo sapiens* for this Catalog was developed by Andrey Davydov; it is not based on any existing domestic or foreign research, methods or theoretical concepts. The laboratory is a partner with the Human Population Academy.

Human Population Academy was founded by Kate Bazilevsky. The Academy's mission is to inform all of over 7 billion humans living on Earth about the discovery of the *Catalog of human population*. The Academy educates about the *Catalog of human population* (*Catalog of Human Souls*) and provides access to informational materials from this Catalog to the public (http://www.humanpopulationacademy.org).

LEADERSHIP

ANDREY DAVYDOV

Research Supervisor of the Special Scientific Info-Analytical Laboratory—Catalog of Human Souls

Andrey Davydov is an expert in Chinese culture, researcher of ancient texts, the author of scientific discovery of the *Catalog of human population* and the technology of decryption of the ancient Chinese monument Shan Hai Jing as the *Catalog of human population*. He authored over 300 published books, including scientific monographs and ideologies. In 2012, he was granted political asylum in the USA due to persecution by a group of employees of the Federal Security Service of Russian Federation (FSB, formerly KGB), who decided to expropriate his research product—the *Catalog of human population*.

OLGA SKORBATYUK

Senior Analyst at the Special Scientific Info-Analytical Laboratory—Catalog of Human Souls

Olga Skorbatyuk is a professional psychologist, one of the developers of the *Catalog of human population*, the founder of Non-Traditional Psychoanalysis, and co-author of over 300 books and scientific articles. She was granted political asylum in the USA together with A. Davydov.

KATE BAZILEVSKY

Founder of the Human Population Academy, Junior Analyst at the Special Scientific Info-Analytical Laboratory—Catalog of Human Souls

Kate Bazilevsky is the director of the Human Population Academy, a Junior Analyst at the Catalog of Human Souls laboratory, an author and a translator of books about the *Catalog of human population*. She holds a degree in MIS and psychology. She founded the Human Population Academy in 2011 and a publishing company called HPA Press in 2012.

www.ingramcontent.com/pod-product-compliance
Lightning Source LLC
Chambersburg PA
CBHW071400290426
44108CB00014B/1626